Photoalignment of Liquid Crystalline Materials

Wiley-SID Series in Display Technology

Series Editor:

Anthony C. Lowe

Consultant Editor:

Michael A. Kriss

Display Systems: Design and Applications
Lindsay W. MacDonald and **Anthony C. Lowe (Eds)**

Electronic Display Measurement: Concepts, Techniques, and Instrumentation
Peter A. Keller

Projection Displays
Edward H. Stupp and **Matthew S. Brennesholtz**

Liquid Crystal Displays: Addressing Schemes and Electro-Optical Effects
Ernst Lueder

Reflective Liquid Crystal Displays
Shin-Tson Wu and **Deng-Ke Yang**

Colour Engineering: Achieving Device Independent Colour
Phil Green and **Lindsay MacDonald (Eds)**

Display Interfaces: Fundamentals and Standards
Robert L. Myers

Digital Image Display: Algorithms and Implementation
Gheorghe Berbecel

Flexible Flat Panel Displays
Gregory Crawford (Ed.)

Polarization Engineering for LCD Projection
Michael G. Robinson, Jianmin Chen, and **Gary D. Sharp**

Fundamentals of Liquid Crystal Devices
Deng-Ke Yang and **Shin-Tson Wu**

Introduction to Microdisplays
David Armitage, Ian Underwood, and **Shin-Tson Wu**

Mobile Displays: Technology and Applications
Achintya K. Bhowmik, Zili Li, and **Philip Bos (Eds)**

Photoalignment of Liquid Crystalline Materials: Physics and Applications
Vladimir G. Chigrinov, Vladimir M. Kozenkov and **Hoi-Sing Kwok**

Photoalignment of Liquid Crystalline Materials: Physics and Applications

Vladimir G. Chigrinov

Hong Kong University of Science and Technology, Hong Kong

Vladimir M. Kozenkov

Moscow State University, Russia

Hoi-Sing Kwok

Hong Kong University of Science and Technology, Hong Kong

A John Wiley & Sons, Ltd., Publication

Telephone (+44) 1243 779777

Email (for orders and customer service enquiries): cs-books@wiley.co.uk
Visit our Home Page on www.wileyeurope.com or www.wiley.com

Other Wiley Editorial Offices

John Wiley & Sons Inc., 111 River Street, Hoboken, NJ 07030, USA
Jossey-Bass, 989 Market Street, San Francisco, CA 94103-1741, USA
Wiley-VCH Verlag GmbH, Boschstr. 12, D-69469 Weinheim, Germany
John Wiley & Sons Australia Ltd, 42 McDougall Street, Milton, Queensland 4064, Australia
John Wiley & Sons (Asia) Pte Ltd, 2 Clementi Loop #02-01, Jin Xing Distripark, Singapore 129809
John Wiley & Sons Canada Ltd, 6045 Freemont Blvd, Mississauga, Ontario, L5R 4J3, Canada

Wiley also publishes its books in a variety of electronic formats. Some content that appears in print may not be available in electronic books.

British Library Cataloguing in Publication Data

A catalogue record for this book is available from the British Library

ISBN 978-0-470-06539-6 (HB)

Typeset in 10/13 Times by Laserwords Private Limited, Chennai, India
Printed and bound in Great Britain by TJ International, Padstow, Cornwall

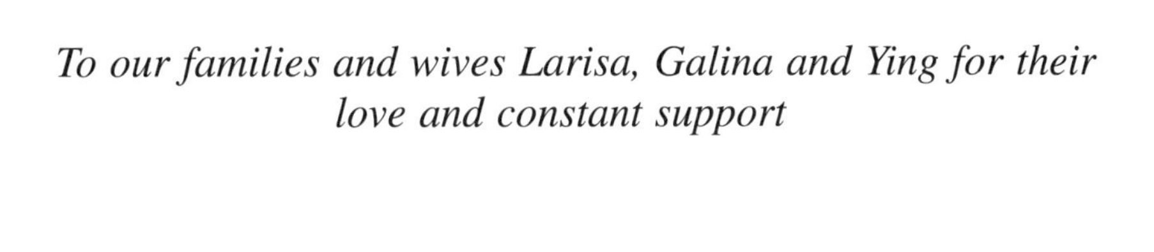

To our families and wives Larisa, Galina and Ying for their love and constant support

Contents

About the Authors

Vladimir G. Chigrinov obtained his PhD degree in solid-state physics from the Shubnikov Institute of Crystallography, USSR Academy of Sciences, in 1978. In 1988, he defended his doctoral degree and became a professor at the Shubnikov Institute of Crystallography, where he was a leading researcher from 1996. He joined HKUST in 1999 and is currently an associate professor. Since 1974 Professor Chigrinov has published 2 books, 15 reviews and book chapters, about 150 journal papers, more than 300 conference presentations, and holds more than 50 patents and patent applications in the field of liquid crystals. He is a Senior SID Member, SID Fellow and an Associate Editor of the *Journal of the Society for Information Display*.

Vladimir M. Kozenkov graduated from the Moscow Energetic Institute as a scientist in applied physical optics (laser department). For 30 years he worked at the Organic Intermediates and Dyes Institute (NIOPIK) in Moscow. He pioneered research and development of various organic photosensitive materials for holography, waveguide, integral and polarization optics, stereolithography, optical memory, imaging processing, and security applications. He was the first to discover the phenomenon of photoinduced birefringence in polyvinyl-cinnamate films in 1977. He has published more than 100 refereed papers and holds more than 50 patents.

Hoi-Sing Kwok obtained his PhD degree in applied physics from Harvard University in 1978. He joined the State University of New York at Buffalo in 1980 and became a full professor in 1985. He joined HKUST in 1992 and is currently Director of the Center for Display Research (*www.cdr.ust.hk*). Professor Kwok has written over 500 refereed publications and holds 40 patents in laser optics and LCD technologies. He is a Fellow of the OSA, the IEEE, and the SID.

Series Editor's Foreword

Manufacture of liquid crystal displays uses some of the most complex and advanced manufacturing techniques. In active matrix displays, TFT circuits are uniformly manufactured at resolution limits of a micron or less on substrates several square metres in area. In most types of liquid crystal displays, liquid crystal molecules are aligned with high precision in directions parallel and perpendicular to the display substrate. This is usually done by a method which has been in use since the discovery of substrate-aligned layers – rubbing.

Rubbing techniques have developed over many decades. Modern rubbing machines are unrecognisable in comparison to the simple methods used in the 1960s around the time of the birth of liquid crystal displays, but they remain as a somewhat anachronistic method when compared to the advances in technique that have been made in most other aspects of liquid crystal display manufacture. Of course, if traditional rubbing methods (and here I mean physical abrasion) worked perfectly, there would be little reason to change them, but as liquid crystal display technology has advanced, an increasing number of performance and yield-related issues have become attributable to rubbing as an alignment technique. Tribologically induced electrostatic discharge damage of the active matrix circuitry has ever been a problem, although it can be reduced by careful selection and control of the rubbing and alignment layer materials. What have become more difficult or even impossible to control are problems related to high pixel density, or to the requirement for multiple alignment domains on each pixel, or to the use of spacers which are deposited on to the display substrate in the gaps between pixels at a point prior in the process to the alignment step. Rubbing, however well controlled, produces some defects and these become more of a problem as pixel size decreases. With multi-domain alignment, the need to carry out multiple rubbing operations and to mask those areas of the surface not being rubbed is costly and reduces yield. The use of rubbing of substrates with deposited spacers produces a region downstream of the spacer where alignment either is absent or is different from that in the surrounding area. This produces disclination defects, which reduce the contrast ratio of the display.

So what other methods are available to the display manufacturer? Much research and development effort has gone into developing liquid crystal technologies that do not require a rubbed alignment layer. The most obvious exemplar of this is the family of multi-domain vertically aligned nematic liquid crystal displays. However, many other liquid crystal technologies require precise directional alignment of the liquid crystal molecules at one or both of the display substrates. Two non-contact methods have been proposed. One uses ion beams, the other light. Both produce alignment which can be superior to that produced by mechanical rubbing. However, ion beam methods are costly because they operate under vacuum. Because they do not require a vacuum, photoalignment methods are lower cost. They are the subject of this latest book in the Wiley-SID series.

The earliest photoalignment methods involved the selective decomposition of surface adsorbed dye molecules by polarised light to produce a residual anisotropic alignment layer. Such methods can produce stable layers which meet the standards required for manufacture, but there remains an element of uncertainty about the very long term stability of such layers, particularly under high illuminance. The authors of this book have developed techniques to achieve photoalignment by thermodynamic reorientation of alignment layers rather than by decomposition.

But photo-techniques are applicable to more than alignment processes. Retardation films are becoming increasingly important as wider viewing angles and higher contrast ratios are demanded, across the entire visible spectrum, in liquid crystal display products. Photopolymerisable liquid crystals offer the opportunity to match with unprecedented precision and compensate switchable liquid crystal layers. Increasing demand for displays with optimised transmissive and reflective properties requires that retardation films and sometimes polarisers need to be patterned at the pixel or subpixel level and photo-techniques provide capabilities that are transferable from development to manufacturing with relative ease.

The authors are all internationally known in their field. They write with clarity and authority. With good reason, they provide a very detailed review of their own work, but they achieve balance by offering a comprehensive review, replete with references, of all work in the field. Theory, experiment and the application of alignment techniques to working devices are all covered in depth. Indeed, the discussion extends beyond displays to photonic devices, where their intricate geometry often makes them unsuitable for use of mechanical alignment methods. Uniquely, a detailed review of granted patents, listed by subject area and assignee, is also provided.

This book will be an invaluable resource to anyone already undertaking or about to undertake research in this field and also to those in industry who wish to develop and apply photoalignment in manufacture.

Anthony Lowe
Series Editor
Braishfield, UK

1

Introduction

Among all the flat panel display technologies, liquid crystal display (LCD) is the most dominant. The manufacture of LCDs is now very mature and done with huge glass substrates measuring over 5 m^2. The availability of such inexpensive high-resolution displays has accelerated the transformation of our society into a display-centric one. However, despite its dominance, LCD is still in need of further improvement. Among other things, light utilization efficiency, cost, and optical performance such as response times of LCD are still not optimal. It is no wonder that much research is still being performed and new LCD modes are still being invented with better properties such as response time or viewing angles.

In this book, we concentrate on one aspect of LCD manufacture, namely that of the alignment surface. In particular, we shall present a comprehensive review of photoalignment technologies. Photoalignment has been proposed and studied for a long time [1–12]. In fact, the subject of light–molecule interactions has been a fascinating subject of research for a long time and is still capturing the imagination of many people. Light is responsible for the delivery of energy as well as phase and polarization information to materials systems. In this particular case, the alignment of the molecules takes place due to a partial ordering of the molecular fragments

Photoalignment of Liquid Crystalline Materials: Physics and Applications
V. Chigrinov, V. Kozenkov and H.-S. Kwok

after a topochemical reaction of a photoselection (Weigert's effect). While the first photo-patterned optical elements, based on polyvinyl-cinnamate films, appeared in 1977 [1], the technology became an LCD one only at the beginning of the 1990s [2–5]. It was soon shown that these materials could provide high-quality alignment of molecules in an LC cell. Over the last 20 years, many improvements and variations have been made for photoalignment. Commercial photoalignment materials are now readily available. Many new applications, in addition to the alignment of LCD, have been proposed and demonstrated. In particular, the application of photoalignment to active optical elements in optical signal processing and communications is currently a hot topic in photonics research.

Photoalignment possesses obvious advantages in comparison with the usually 'rubbing' treatment of the substrates of LCD cells. Possible benefits for using this technique include:

1. Elimination of electrostatic charges and impurities as well as mechanical damage of the surface.
2. A controllable pretilt angle and anchoring energy of the LC cell, as well as its high thermal and UV stability and ionic purity.
3. The possibility to produce structures with the required LC director alignment within the selected areas of the cell, thus allowing pixel dividing to enable new special LC device configurations for transflective, multidomain, 3D, and other new display types.
4. A potential increase of manufacturing yield, especially in LCDs with active matrix addressing, where the pixels of a high-resolution LCD screen are driven by thin film transistors on a silicon substrate.
5. New advanced applications of LCs in fiber communications, optical data processing, holography, and other fields, where the traditional rubbing LC alignment is not possible due to the sophisticated geometry of the LC cell and/or high spatial resolution of the processing system.
6. The ability for efficient LC alignment on curved and flexible substrates.
7. Manufacturing of new optical elements for LC technology, such as patterned polarizers and phase retarders, tunable optical filters, polarization non-sensitive optical lenses, with voltage-controllable focal distance etc.

With all these new developments in photoalignment technologies, it is now the right time to take an inventory of the progress made over the last 20 years in the form of a monograph. This book presents the status of the research in LCD photoalignment and photo-patterning. To the best of our knowledge, there are no other books devoted to the subject, though a few review articles are available [6–11]. In this book, we shall concentrate on a recent approach of ours, which is rather promising, namely the photoinduced reorientation of azo-dyes [12]. This technique of photoalignment does not involve any photochemical or structural transformations of the molecules. Further, the new photoaligning films are robust and possess rather good aligning properties such as anchoring energies and voltage holding ratios. They can be very useful for the new generation of LC devices as well as in new photovoltaic, optoelectronic, and photonic devices based on highly ordered thin organic layers. Examples of such applications are light-emitting diodes (OLEDs), solar cells, optical data storage, and holographic memory devices. The novel and highly ordered layer structures of organic molecules may exhibit certain physical properties, which are similar to the aligned LC layers.

This book is intended for a wide audience including engineers, scientists, and managers, who wish to understand the physical origins of the photoaligning technique, its basic advantages and limitations, as well as the application for LC devices, including displays, optical waveguides, optical polarizers and retarders, etc. University researchers and students who specialize in condensed matter physics and LC device development should also find some useful information here.

The principal aims of the book are:

1. To describe the physical mechanisms of LC photoalignment with a special emphasis on the most useful photosensitive materials and preparation procedures suitable for the purpose.
2. To summarize LC surface interaction in photoaligned LC cells to produce the required LC pretilt angles, anchoring energy, ionic purity, IR and UV stability, and sensitivity to the activating light exposure.
3. To show how to produce perfect vertical, twisted, rewritable, and other LC photoalignment in nematic, ferroelectric, lyotropic, and discotic materials on glass and plastic substrates and special LC cell configurations (Si waveguide and 3D surfaces, superthin tubes, etc.).
4. To compare various applications of photoalignment technology for in-cell patterned polarizers and phase retarders, transflective and microdisplays, security and other LC devices.

5. To present recent results in applications of photoaligning and photo-patterning technology in LC devices.

The organization of this book is as follows. In Chapter 2, we shall present the various photoalignment mechanisms. Here the photoaligning techniques will be described and compared. In Chapter 3, the alignment properties of the various films will be discussed, with an emphasis on the azo-dyes. Properties such as anchoring energies and voltage holding ratios are important to LCD applications and will be discussed in detail. In Chapter 4, the application of photoalignment to various LCD modes and LC applications will be presented. Photoalignment on unconventional substrates will also be described to illustrate the power of such a technique. In Chapter 5, various applications of photoalignment to the fabrication of optical elements will be explored. The use of photoalignment to improve the design of transflective displays will be presented as well. Some applications of photoaligning technology for the development of new LC displays and photonic devices are then listed in Chapter 6. The working prototypes of new photoaligned LC devices, e.g. optically rewritable electronic paper, look very promising for future applications. One special feature of the book is a compilation of the most important patents, which forms Chapter 7. They are also classified in various ways for easy comprehension of where the technology is heading. We believe such a compilation will be very useful to readers.

The topic of this book is of particular interest to us, as we have undertaken some of the pioneering research in the field. In a sense it is a form of stock taking for us. We hope that the book will stimulate new research and development in the field of LC photoalignment and enable the technology to be used in large-scale LCD production. We are grateful to many colleagues who have worked with us in the past and are still working with us at the Center for Display Research in the Hong Kong University of Science and Technology, including E. P. Pozhidaev, W. C. Yip, Fion Yeung, Jacob Ho, Y. W. Li, A. Muravsky, A. Murauski, O. Yaroshchuk, A. Kiselev, V. Shibaev, S. A. Pikin, A. Verevochnikov, E. Prudnikova, V. Konovalov, S. Pasechnik, Z. H. Ling, D. D. Huang, X. Li, P. Xu, G. Hegde, and H. Y. Mak. They have provided much important information and have contributed greatly to our research program. We also owe much to our friends Drs H. Takatsu, H. Takada, H. Hasebe, and M. Schadt for many stimulating discussions.

References

[1] E. D. Kvasnikov, V. M. Kozenkov, and V. A. Barachevskii, Birefringence in polyvinyl cinnamate films, induced by polarized light. *Dokladi Akademii Nauk SSSR* **237**, 633 (1977) (in Russian).

[2] K. Ichimura, Y. Suzuki, T. Seki, A. Hosoki, and K. Aoki, Reversible change in alignment mode of nematic liquid crystals regulated photochemically by 'command surfaces' modified with an azobenzene monolayer. *Langmuir* **4**, 1214 (1988).

[3] W. M. Gibbons, P. J. Shannon, S. T. Sun, and B. J. Swetlin, Surface mediated alignment of nematic liquid crystals with polarized laser light. *Nature* **351**, 49 (1991).

[4] M. Schadt, K. Schmitt, V. Kozenkov, and V. Chigrinov, Surface-induced parallel alignment of liquid crystals by linearly polymerized photopolymers. *Japanese Journal of Applied Physics* **31**, 2155 (1992).

[5] A. G. Dyadyusha, T. Ya. Marusii, Yu. A. Reznikov, A. I. Khizhnyak, and V. Yu. Reshetnyak, Orientational effect due to a change in the anisotropy of the interaction between a liquid crystal and a bounding surface. *JETP Letters* **56**, 17 (1992).

[6] L. M. Blinov, Photoinduced molecular reorientation in polymers, Langmuir Blodgett films and liquid crystals. *Journal of Nonlinear Optical Physics and Materials* **5**, 165 (1996).

[7] M. O'Neill and S. M. Kelly, Photoinduced surface alignment for liquid crystal displays. *Journal of Physics D: Applied Physics* **33**, R67 (2000).

[8] M. Schadt, Liquid crystal displays and novel optical thin films enabled by photoalignment. *Molecular Crystals and Liquid Crystals* **364**, 151 (2001).

[9] V. G. Chigrinov, V. M. Kozenkov, and H.-S. Kwok, New developments in photoaligning and photo-patterning technologies: physics and applications. In *Optical Applications of Liquid Crystals*, Ed. L. Vicari, Institute of Physics Publishing, Bristol and Philadelphia, 201–244 (2003).

[10] V. P. Shibaev, S. A. Kostromin, and S. A. Ivanov, *Polymers as Electrooptical and Photooptical Active Media*, Springer-Verlag, Berlin, 37–110 (1996).

[11] V. Shibaev, A. Bobrovsky, and N. Boiko, Photoactive liquid crystalline polymer systems with light-controllable structure and optical properties. *Progress in Polymer Science* **28**, 729 (2003).

[12] V. Chigrinov, S. Pikin, A. Verevochnikov, V. Kozenkov, M. Khazimullin, J. Ho, D. Huang, and H.-S. Kwok, Diffusion model of photoaligning in azo-dye layers. *Physical Review E* **69**, 061713 (2004).

2

Mechanisms of LC Photoalignment

The effect of LC photoaligning is a direct consequence of the appearance of the photoinduced optical anisotropy and dichroic absorption in thin amorphous films, formed by molecular units with anisotropic absorption properties [1]. The first publication on LC photoalignment appeared in 1988, which discussed the application of the reversible *cis-trans* isomerization of azobenzene molecular layers attached to a solid surface to the switching of the alignment of the adjacent LC layer from homeotropic to the azimuthally random planar orientation [2]. Optical control of LC alignment was achieved by changing the wavelength of the non-polarized light illumination [2]. Later it was shown that the alignment of an LC medium could be achieved by illuminating a dye-doped polymer alignment layer with polarized light [3]. LC molecules in contact with the illuminated area were homogeneously aligned perpendicular to the direction of the laser polarization and remained aligned in the absence of the laser light. Soon after the LC photoalignment procedure was derived using cinnamoyl side-chain polymers [4, 5] and polyimide aligning agents [6]. The area of LC photoalignment is developing

Photoalignment of Liquid Crystalline Materials: Physics and Applications
V. Chigrinov, V. Kozenkov and H.-S. Kwok

very rapidly now and the vast majority of new materials, techniques, and LCD prototypes based on photoalignment technology have appeared only recently and are partially covered by some recent review articles [7–10].

Progress in the application of LC photoaligning for LCDs has stimulated many fundamental studies of its mechanism. However, an adequate explanation of the photoalignment phenomenon is still absent. Several publications have appeared recently and provide only a qualitative explanation of the phenomenon of photoalignment. A more adequate explanation of the process is still in progress, taking into account both photochemical and non-photochemical transformations in photosensitive layers under the action of polarized or non-polarized, but directed, light.

We can distinguish the following mechanisms of the photoalignment: (i) photochemical reversible *cis-trans* isomerization in azo-dyes containing polymers [11–16], monolayers [7, 17], and pure dye films [2, 3]; (ii) pure reorientation of the azo-dye chromophore molecules or azo-dye molecular solvates due to diffusion under the action of polarized light [18]; (iii) topochemical crosslinking in cinnamoyl side-chain polymers [4, 5]; (iv) photodegradation in polyimide materials [6, 19–21]. The method of repeated *cis-trans* photoisomerization reaction resulting in the reorientation of the backbone structure of polyamic-acid-contained azobenzene units has also been reported [22, 23]. The subsequent imidization stabilized the polyimide structure; the LC alignment was thermally and optically stable and no decomposition was involved. Bare solid surfaces with adsorbed organic molecules have been used to orient LCs after UV light irradiation [24]. The detailed structure and behavior of the surface-adsorbed molecules are not important to the effect: just their UV light absorption should depend on their orientation. The only requirement of the solid substrates is their transparency to the UV light. The universal reason for the photoinduced anisotropy in such systems is that photons destroy on the surface those molecules that absorb them most intensively. This is equivalent to a kind of light rubbing resulting in the anisotropic ablation of the adsorbed material.

Processes (i) and (iv) involving azo-dyes present reversible transformations, while the other two processes require irreversible photochemical changes. For these latter processes, since chemical changes occur, it is difficult to maintain the purity of the alignment layer. For *cis-trans* isomerization the change of the absorption spectra is observed after illumination, which is not the case for the photochemically stable azo-dye molecules involved in a reorientation and solvate formation process under the action of polarized light [18, 25]. These two processes provide a high purity of the aligning layers suitable for LCD applications with thin film transistors (TFT-LCD). Let us consider the mechanisms in more detail and illustrate them with various photoaligning materials.

2.1 Cis-Trans Isomerization

2.1.1 'Command Surface'

Under the action of polarized light a reversible *trans-cis* transformation of azo-dye molecules can be observed [2]. Ichimura *et al.* showed that, if the dye molecules are directly attached to the surface, than a so-called 'command surface' can be created [2, 26–30] (Figure 2.1).

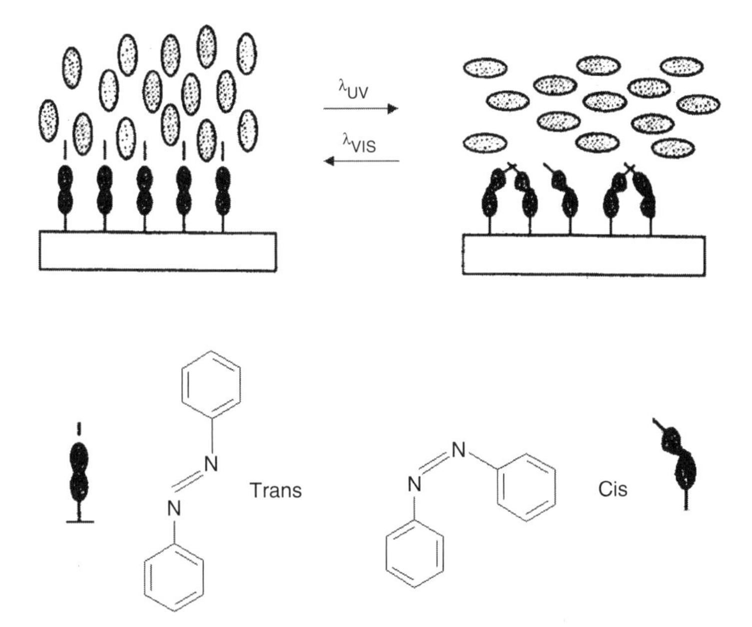

Figure 2.1 Reversible homogeneous to homeotropic transition in LC cell, caused by *cis-trans* isomerization of the azobenzene units, attached at the glass substrate ('command surface'). UV light ($\lambda = 365$ nm) transforms the azobenzene units to the *cis* form (homogeneous LC alignment), while visible light ($\lambda = 440$ nm) restores the *trans* form configuration [2]. Reproduced from [2], (1988), American Chemical Society

The conformation of the azobenzene dye units was shown to undergo the reversible *cis-trans* transformations under subsequent illumination by UV–visible non-polarized light. Thus the ensemble of the LC molecules will reversibly change its orientation from homeotropic to random planar and vice versa, so that one surface dye unit may cause a reorientation of about a million LC molecules in the bulk. Such a surface with a reversible *cis-trans* isomerization taking place under the action of UV–visible light transformation was called a 'command surface' [2]. However, this method of controlling the LC reorientation reveals certain disadvantages, such as: (i) the homeotropic to planar configuration is azimuthally degenerate; (ii) the dark relaxation of the *cis* configuration is observed, thus spontaneously restoring the initial homeotropic LC configuration; and (iii) the number of reversible cycles is limited by a photochemical stability of the dye layers.

Ichimura *et al.* have shown that the effect of *cis-trans* isomerization is very important in many cases [2, 26, 27]. The application of photo-optical regulation of LC orientation in waveguide structures [28], photoaddressing, and photorecording [29] has been proposed. The photoinduced anchoring from homeotropic to random planar and uniform planar alignment has been studied depending on the structure of the photoisomerized azo-dyes [30]. The new silane [31] and stilbene [32] surfactants were used to produce the photoregulated alignment.

2.1.2 Cis-Trans Transformations in Azo-Dye Side-Chain Polymers and Azo-Dye in a Polymer Matrix

The concept of *cis-trans* isomerization in azo-dye side-chain polymers was developed by Shibaev *et al.* [11]. The side chain of the polymer was made of azobenzene side groups (Figure 2.2). After illumination with UV light the following transformation occurs [33] (Figure 2.2):

$$\textit{Trans} \text{ (parallel to the UV light polarization)} \Rightarrow \textit{Cis}$$

$$\Rightarrow \textit{Trans} \text{ (perpendicular to the UV light polarization).}$$

The side-chain polymers used by Shibaev *et al.* are also shown in Figure 2.2. The increase of the molar percentage of the azobenzene moiety results in a higher order of the azo-dye fragments. The order parameter of the azo-dye side-chain polymers cannot be very high, however, as it can be considerably increased when the annealing of the photo-oriented azo-dye side-chain polymer is used above the glass transition temperature [34]. Theoretical approaches that describe *cis-trans*

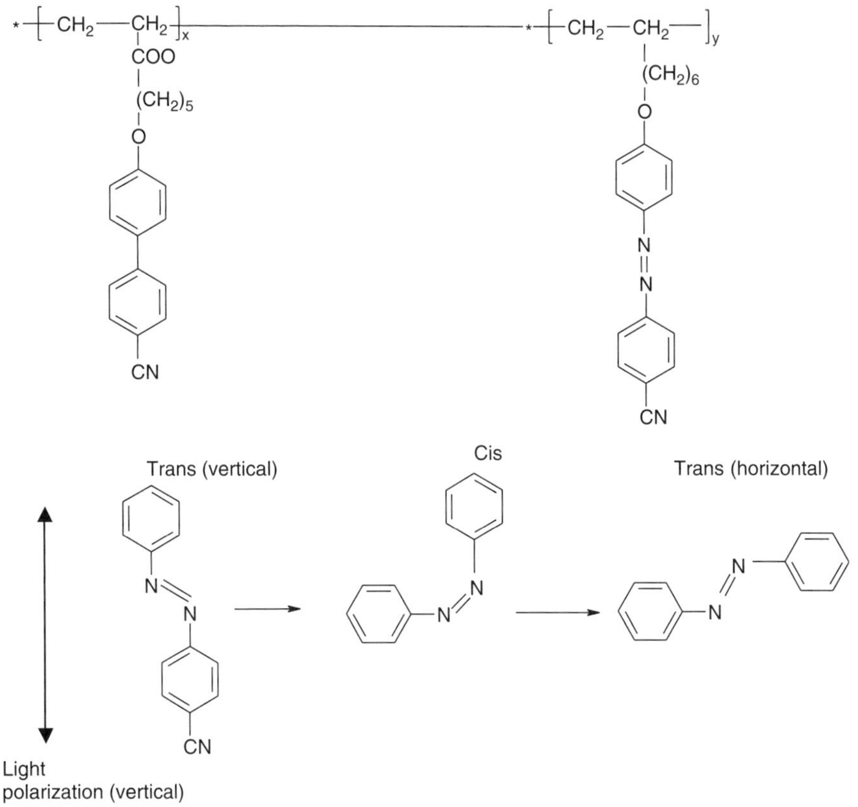

Figure 2.2 Side-chain polymers, used by Shibaev *et al.* [11], which show the effect of *cis-trans* isomerization in a field of polarized UV light. The *trans* isomer, which is parallel to the UV light polarization vector, is transferred to the *cis* isomer and then again to the *trans* isomer, which is perpendicular to the initial one. As a result of this transformation, all the absorption oscillators of azobenzene side-chain molecules align perpendicular to the UV light polarization. Reprinted from [10], Institute of Physics Publishing (2003)

isomerization in azobenzene LC side-chain polymers are provided in [12–16]. Yaroshchuk *et al.* have investigated photoinduced anisotropy and the order parameter in azopolymer films and found that the order parameter exceeds that of Langmuir films [35]. The 3D photo-orientation of azochromophores in azobenzene LC side-chain polymers is dependent on their chemical structure and the method of photoalignment was carefully studied by the method of attenuated total internal reflection (ATR) [36] and transmission null ellipsometry (TNE) [16, 37, 38].

Biaxial, uniaxial, as well as isotropic 3D ordering were detected for various orientations of azochromophores and classified by analogy to crystal optics as positive and negative C and A plates [36].

The process shown in Figure 2.2 has been further observed by Gibbons *et al.*, who dissolved diazodiamine molecules in a polymer matrix of silicone polyimide copolymer [3]. After a spin-coating procedure, the homogeneous LC cell was prepared by rubbing the polymer layer. Then the cell was illuminated by Ar^+ laser light near the wavelength of the maximum absorption of the dye ($\lambda \approx 490$ nm) with the direction of the illumination parallel to the rubbing axis (Figure 2.3).

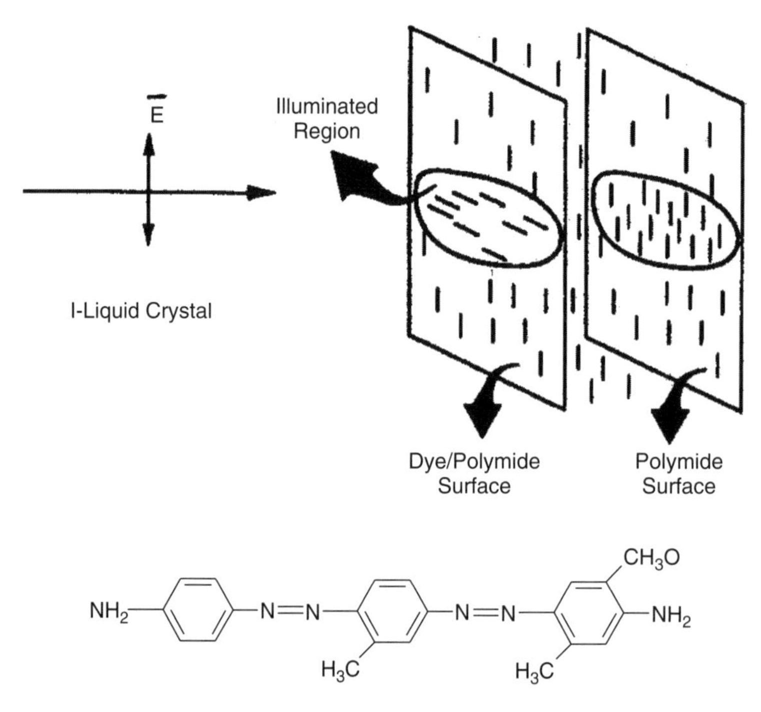

Figure 2.3 The aligning of the LC cell using the effect of rotation of the azo-dye absorption oscillator perpendicular to the activating light polarization vector. The structure of the used diazodiamine dye is shown below. LC molecules are aligned parallel to the azo-dye molecules within the illuminated region, thus forming a twist alignment from the initial homogeneous configuration [3]. Reproduced from W. M. Gibbons, P. J. Shannon, S. T. Sun, and B. J. Swetlin, "Surface mediated alignment of nematic liquid crystals with polarized laser light." *Nature* **351**, 49 (1991), Nature Publishing Group

Within the illuminated region, the azo-dye molecules were aligned perpendicular to the direction of light polarization, thus resulting in a twist alignment of the LC cell [3]. Subsequently, Gibbons *et al.* observed the aligning effect of azo-dyes in LC volumes, using a 'guest–host' mixture [39]. Later Gibbons *et al.* used an LC cell of 40–50 μm thickness with initial homogeneous alignment on an azo-dye/polyimide-coated substrate to record grayscale images, produced by Ar^+ laser light with a wavelength of 514.5 nm [40, 41]. The reorientation of the azo-dye molecules perpendicular to the polarization of the UV light caused the subsequent reorientation of the neighboring LC layer, thus increasing the contrast of the optical image.

2.2 Pure Reorientation of the Azo-Dye Chromophore Molecules or Azo-Dye Molecular Solvates

Another mechanism is related to the reorientation of the dye molecules due to the action of the polarized light illumination [42, 43]. The reorientation of the azo-dye molecules perpendicular to the polarization of the activated light was first observed by Kozenkov *et al.* in Langmuir–Blodgett (LB) films [44, 45]. The LB films, made of azo-dyes, were transferred from the surface of water to polished quartz glass. The induced value of optical anisotropy measured at the wavelength $\lambda =$ 632.8 nm was $\Delta n \cong 0.23$. The dichroic ratio at the wavelength of the maximum absorption of the dye, $\lambda = 500$ nm, was $N = D_{\perp}/D_{\parallel} \geq 4.6$, which corresponds to the order parameter $S = (1 - N)/(1 + 2N) = -0.35$ (70% of the dye molecules are arranged perpendicular to the light polarization vector [46]).

Besides a chemical reaction, the UV light can induce an asymmetric potential field under which the stable configuration is characterized by the dye absorption oscillator perpendicular to the induced light polarization. One of the possible photoaligning mechanisms in azo-dye films is a pure reorientation of azo-dye molecules [10, 18, 42, 43]. Diffusion mechanisms of the azo-dye reorientation can be used to explain this phenomenon.

2.2.1 Diffusion Model

Basic concept

When the azo-dye molecules are optically pumped by a polarized light beam, the probability of absorption occurring is proportional to $\cos^2\theta$, where θ is the angle

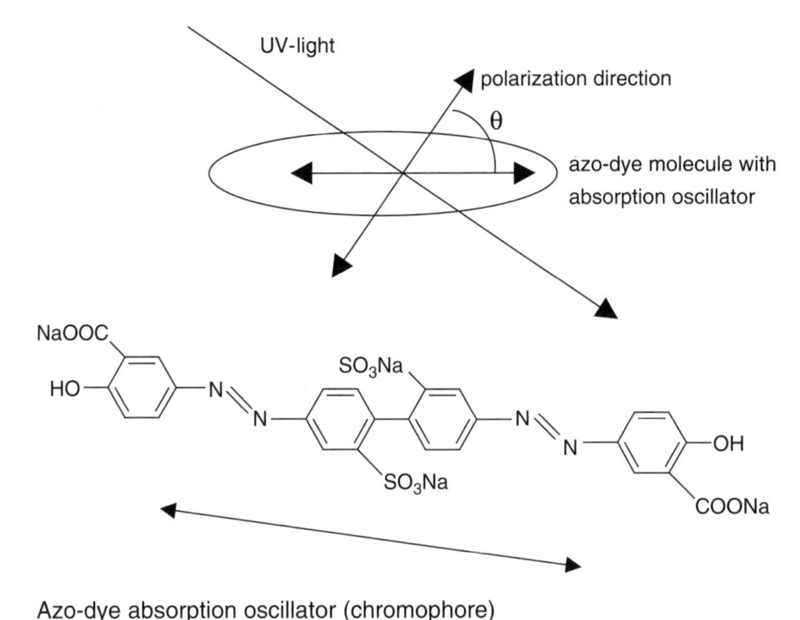

Figure 2.4 The qualitative interpretation of the photoinduced order in photochemical stable azo-dye films: upper, the geometry of the effect; lower, the azo-dye molecule, having the absorption oscillator (chromophore) perpendicular to the long molecular axis [10, 18, 47]

between the absorption oscillator of the azo-dye molecules and the polarization direction of the light (Figure 2.4).

Thus the azo-dye molecules, which have their absorption oscillators (chromophores) parallel to the light polarization, will most probably get an increase in energy which results in their reorientation from their initial position. This results in an excess of chromophores in a direction where the absorption oscillator is perpendicular to the polarization of the light.

We have studied this photoinduced reorientation effect carefully. In our case the chromophore is parallel to the long molecular axis of the azo-dye (Figure 2.4), i.e. azo-dye molecules are tending to align their long axes perpendicular to the UV light polarization.

For rod-like azo-dye molecules, shown in Figure 2.4 (sulfonic azo-dye SD1), with cylindrical symmetry the only coordinate will be the polar angle θ, the angle between the molecular absorption oscillator and the direction of polarization of the activating light. The potential energy $U = \Phi kT$, where T is the absolute temperature, k Boltzmann's constant, and Φ the relative potential energy of the activating light that will also be a function of the angle θ only, i.e. $\Phi = \Phi(\theta)$, and thus the equation of orientational diffusion can be written as [18]

$$\frac{\partial^2 f}{\partial \theta^2} + \frac{\partial}{\partial \theta}\left(f \frac{\partial \Phi}{\partial \theta}\right) = \frac{1}{D}\frac{\partial f}{\partial t}, \tag{2.1}$$

$$\Phi = \frac{A}{2}\cos^2\theta, \quad A = \frac{I\alpha V_M \tau}{kT}. \tag{2.2}$$

Here we assume that the potential energy ΦkT is proportional to the light intensity in the direction of the absorption oscillator, i.e. the number of photons absorbed by the azo-dye molecule. The energy of photons is not wasted in the change of the molecular conformation or luminescence and is responsible for the rotation of the azo-dye molecule from its initial random position to the direction perpendicular to the activating light polarization [10].

The order parameter S is determined as the thermodynamic average $S = \langle P_2(\theta)\rangle$:

$$P_2(\theta) = \frac{1}{2}(3\cos^2\theta - 1),$$
$$S = \langle P_2 \rangle = \int_0^{\pi} P_2(\theta) f(\theta, t)\sin\theta \, d\theta \Big/ \int_0^{\pi} f(\theta, t)\sin\theta \, d\theta. \tag{2.3}$$

Here, the probability density function $f(\theta, t)$ describes the movement of molecules in two dimensions, i.e. in a certain plane, where the angle θ characterizes the orientation of the dye molecule with respect to the polarization vector of the activating light. The orientation of the azo-dye molecule in the perpendicular plane is not controlled by light. The value $f(\theta, t)\,d\theta$ is the number of molecules whose axes are located within the polar angles $(\theta, \theta + d\theta)$ at the unit sphere [18].

Equations (2.1) and (2.2) are correct if there is no interaction between azo-dye molecules. Such a situation can occur in very dilute systems with azo-dyes, e.g. in special polymer films having a small amount of azo-dye molecules. Depending on the sign of the parameter A, the molecular orientation occurs either at the angle $\theta = \pi/2 (A > 0)$ or at $\theta = 0, \pi (A < 0)$. Equation (2.2) includes also the light power $I(\mathrm{W/cm^2})$, the absorption coefficient $\alpha(\mathrm{cm^{-1}})$, and the molecular volume V_M, absolute temperature T and Boltzmann's constant k. The parameter τ is the relaxation time – the time of 'cooling' of the azo-dye molecule and its surroundings. During this time τ the azo-dye molecule absorbs the energy of the light, overcomes the potential barrier between two orientational states when it interacts with its surroundings, and then 'cools down' to the new state. The parameter $U = \Phi kT = \frac{1}{2} I\alpha V_M \tau \cos^2\theta$ plays the role of a certain potential which makes the azo-dye molecule rotate due to the action of the light, and can be considered as the change of chemical potential per molecule related to the corresponding change of the free energy. The process of such a reorientation can be considered as being isothermal. During this process the free energy of the azo-dye molecules changes due to the action of the light, and the thermal energy becomes scattered. We should point out that the change of the free energy occurs only for the reoriented molecules. After this reorientation the molecules are not affected by light again, i.e. this free energy change per molecule is a constant value. Only the quantity of reoriented molecules increases with time, and this is described by the variation of the angular distribution function f.

Unfortunately, the parameter τ cannot be exactly calculated due to the complicated nature of the numbered interactions and molecular kinetics, but τ should be much smaller than the time of light exposure to enable the azo-dye molecules to get to a new orientational state. Parameter τ can be estimated from experiments on photoaligning. If $kT \sim 10^{-20}$ J (room temperature), $\alpha \sim 10^5$ cm^{-1}, $I \sim 10^{-2}$ W/cm^2, $V_M \sim 10^{-20}$ cm^3, then $\tau \sim 10^4$ to 10^{-3} s for $A \sim 0.1$ to 1. The latter limits of the parameter values have been found from experimental data [10, 18]. We assume that the characteristic molecular size is about 20–30 angstroms (Figure 2.4) and estimate the absorption coefficient from the measured optical density $D (D \sim 0.2$ for an azo-dye layer thickness of 10 nm) [10]. The values of the coefficient $A \sim 0.1 - 1$ provide the best fitting of the experimental curves by the diffusion model (see the results below). The time τ may be dependent on the thermal conductivity of substance κ, heat capacity c_p, density ρ, and characteristic size of sample L on which the distribution of temperature becomes homogeneous, i.e. $\tau \approx L^2 c_p \rho/\kappa$. Since $\rho \approx 1$ g/cm^3, $c_p \approx 1.6$ J/g/K, $\kappa \approx 0.1$ W/m/K, we obtain the estimate $\tau \sim 10^{-4}$ s for $L \approx 2$ μm [18].

The function $f(\theta)$ of the statistical distribution of the azo-dye molecule aligning along the various orientations θ, which is $f = \pi/4$ in the initial state, will tend to $f = \delta(\theta - \pi/2)$ for a sufficiently high exposure time (Figure 2.5).

Hence, a thermodynamic equilibrium in the new oriented state will be established. Consequently, the anisotropic dichroism or birefringence is photoinduced permanently and the associated order parameter as a measure of this effect goes to a saturation value, which can be very large in these materials.

The perfect order $\theta \equiv \pi/2$ occurs when the absolute value of the order parameter is maximal, i.e. $S_m = -1/2$ (Equation (2.3)). Then we can define the value of the relative order parameter $s = S/S_m (0 \leq s \leq 1)$ [18]. The relative order parameter

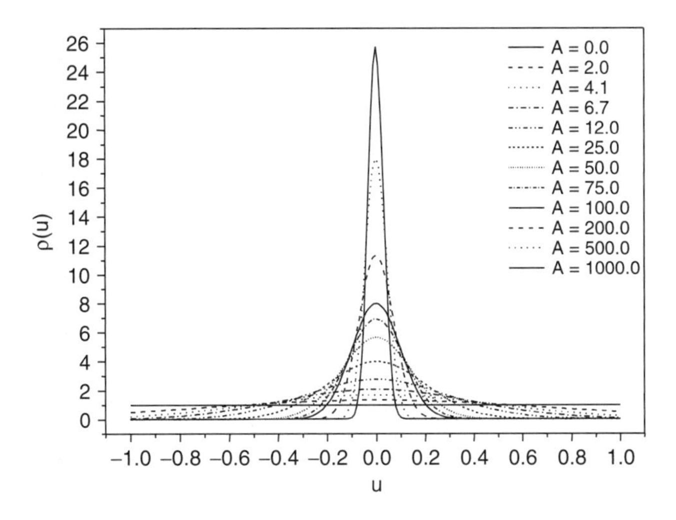

Figure 2.5 The distribution function. $\rho(u), u = \cos\theta$, of the azo-dye molecules for various values of the parameter A, proportional to the intensity of the activated light (2.2)

for small values of A is [18]

$$s \approx \frac{2A}{15}\left[1 - \exp\left(-\frac{t}{\tau_{rise}}\right)\right] \tag{2.4}$$

or the saturation value of the order parameter is directly proportional to the intensity I of the activating light (2.2). The response (rise) time is

$$\tau_{rise} = \frac{1}{6D} \approx \frac{\pi\eta a_m}{kT}, \tag{2.5}$$

where $D = kt/(6\pi\eta a_m)$ is a diffusion coefficient, η the viscosity, and a_m the characteristic size of the molecule or molecular cluster. Our experimental data clearly show that the azo-dye SD1 clusters are very uniformly distributed in the surface area of the substrate, while the cluster size does not depend on the exposure time, as is shown in the atomic force microscope (AFM) picture of SD1 with a thickness of 20 nm on top of an indium tin oxide (ITO) film in Figure 2.6. Relation (2.5) states that the rise time of the effect is independent of the intensity of the activating light.

Looking at the kinetics of the process for the induced birefringence, we get

$$\left.\frac{ds}{dt}\right|_{t=0} \approx \frac{2A}{15\tau_{rise}} = \frac{4AD}{5}, \tag{2.6}$$

Figure 2.6 AFM picture of the azo-dye layer SD1 on a glass substrate

i.e. the rate of the induced order parameter $ds/dt|_{t=0}$ is directly proportional to the relative intensity of the activated light.

In the general case, particularly for dye films, we should take into account the interaction between dye molecules themselves in the process for the rotational diffusion. We can use the approximation of the mean field acting as the thermodynamic average $\langle P_2(\theta)\rangle$. Thus the effective potential, at which the dye molecule stays, can be written in the form

$$U = \frac{1}{2} I\alpha V_M \tau \cos^2\theta + a\langle P_2\rangle P_2(\theta), \tag{2.7}$$

where α is a phenomenological constant. The potential (2.7) is the 'mean field' approximation, similar to that proposed in the Maier–Saupe model [46]. The amplitude of the mean field is proportional to the order parameter $S = \langle P_2\rangle$, while the function $P_2(\theta)$ makes the interaction potential orientation dependent [18].

We shall assume, first, that the potential U in the equation of diffusion is given by Equation (2.7), and the initial function $f_0(\theta) = 1/2$ is isotropic. Thus taking into account (2.3) and (2.7), the equation for the function $f(\theta, t)$ can be written in the form

$$\frac{\partial^2 f}{\partial \theta^2} - \left(\frac{1}{2}A + \frac{3}{2}\frac{a}{kT}\langle P_2\rangle\right)\frac{\partial}{\partial \theta}[f\sin(2\theta)] = \frac{1}{D}\frac{\partial f}{\partial t}. \quad (2.8)$$

The solution of (2.8) can only be found numerically in the whole time interval from $t = 0$, when $f(\theta, 0) = 1/2$ and $S = 0$, to $t \to \infty$, when certain limits for f and S have to be observed. When the value of the relative exposure energy A in (2.2) is not small and the initial isotropic angular distribution function $f_0 = 1/2$ is highly distorted, we have to use numerical methods to find the solutions of the diffusion equation (2.1). One of the interesting possibilities here is to consider the molecular interaction potential in the form

$$\Phi = \frac{1}{2}A\cos^2\theta + \frac{a}{kT}\langle P_2\rangle P_2(\theta), \quad (2.9)$$

$$\Phi = \frac{a}{kT}\langle P_2\rangle^f P_2(\theta) \quad (2.10)$$

where (2.9) and (2.10) correspond to the exposure process and relaxation respectively and $\langle P_2\rangle^f$ is the 'frozen' order parameter during the relaxation process $(A = 0)$,

$$\langle P\rangle^f = \langle P_2\rangle^{off} = \langle P_2\rangle\left(t_{off}\right) \quad (2.11)$$

and $t_{exp} = t_{off}$ – exposure time. This means that the coefficient $(a/kT)\langle P_2\rangle^f$ before the term P_2 in (2.9) and (2.10) remains constant during the relaxation of the order parameter. The saturated value of the order parameter $\langle P_2\rangle|_{t\to\infty}$ increases with the exposure time, as the coefficient $(a/kT)\langle P_2\rangle^f$ becomes larger for higher values of τ_{exp}. The situation is confirmed by the experiment described below.

Experimental

The purpose of our experimental investigations was to measure UV-light-induced birefringence which is proportional to the order parameter of a photoaligning substance. The photosensitive sulfonic azo-dye SD1 was used in the experiment (Figure 2.4). The azo-dye was dissolved in N,N-dimethylformamide at a concentration of 1 wt% and then spin coated on an ITO-coated glass substrate at 3000 rpm for 30 s. The coated substrate was cured for 15 min at $T = 145\,^\circ$C. This procedure

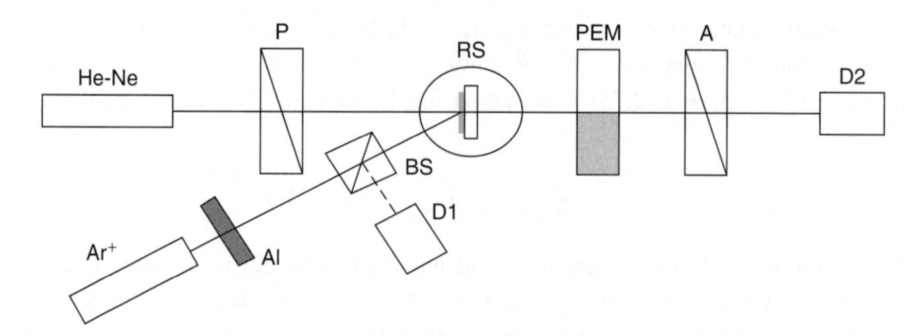

Figure 2.7 Experimental setup: PEM, photoelastic modulator; P, polarizer; A, analyzer; RS, rotatory stage with SD film; AT, attenuator; D1 and D2, detectors; Ar^+ and He–Ne, lasers; BS, beam splitter

provides a thin homogeneous solid dye film on an ITO glass surface. For accurate *in situ* measurement of a small thin film phase retardation δ, the experimental setup shown in Figure 2.7 was constructed [18].

The photoelastic modulator enabled the phase retardation value to be measured with an accuracy of ± 0.001 rad. The SD1 birefringence was measured *in situ* during the UV irradiation process. The pump beam for inducing optical anisotropy in the SD film was provided by the Ar^+ laser ($\lambda = 457$ nm), which outputs linear polarized light at a power $W = 10$ mW/cm^2. The output power was controlled by a special set of attenuator plates (AT). The probe beam, from the He–Ne laser ($\lambda = 632.8$ nm), passes through the crossed Glan–Thompson polarizers (P and A), the substrate with the SD film mounted on a rotatory stage (RS), and the photoelastic modulator (PEM). The transmitted probe light intensity I was modulated by the PEM at a frequency $\Omega = 50$ kHz and detected by the semiconducting photodetector (D2).

The photo-induced phase retardation $\delta = 2\pi \Delta n d/\lambda$ (azo-dye layer thickness $d \approx 10$ nm, $\lambda = 632.8$ nm) measured in the experiment is directly proportional to the relative order parameter s of the azo-dye layer, i.e.

$$\delta = \frac{2\pi \Delta n d}{\lambda} = ks, \tag{2.12}$$

where k is a coefficient proportional to the anisotropy of the molecular polarizability of the azo-dye molecule [46].

The dependence of the phase retardation versus exposure time for different powers W of the pumped Ar^+ laser is shown in Figure 2.8.

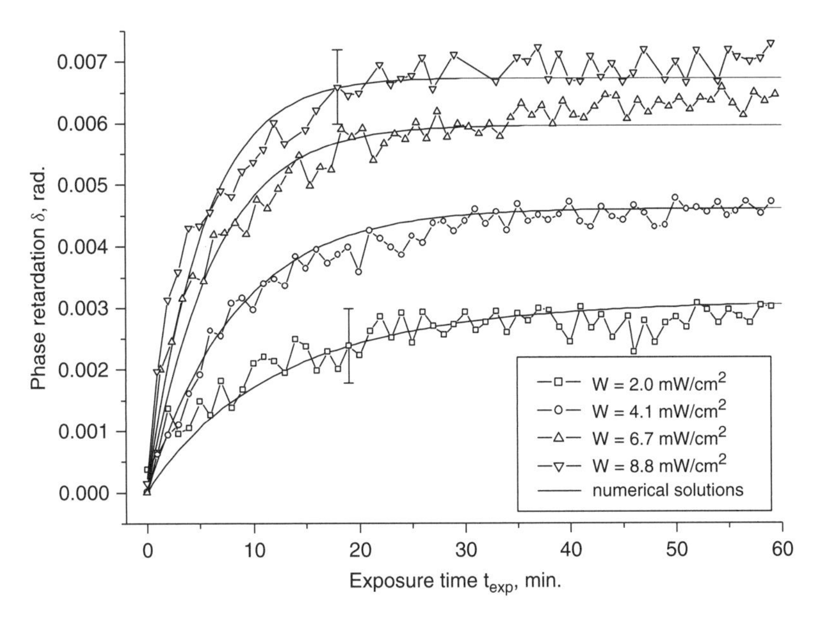

Figure 2.8 Dependence of the photoinduced phase retardation δ on exposure time t_{exp} for various powers W of the activated UV light [18]. Solid lines indicate the numerical solution of the diffusion model (2.1)–(2.2)

The difference in phase retardation did not exceed the experimental error of ± 0.001 rad. A monotonic increase of the photoinduced phase retardation δ from zero level at $t = 0$ (isotropic state) up to the saturation value δ_0 (anisotropic state) was observed. The experimental values are in good agreement with the results of numerical calculations, using diffusion equation (2.1), shown in Figure 2.8 by solid lines. The saturation photoinduced phase retardation level was found to be proportional to the power of UV illumination W, which is in agreement with the results of numerical calculations (see Figure 2.9).

The initial rate of increase of the photoinduced birefringence should also establish a linear dependence as in Equation (2.6) (see Figure 2.10).

The derivatives $(d\delta/dt)_{t=0}$ were calculated from experimental data (Figure 2.8). The numerical results, as follows from the model of orientational diffusion of azo-dye molecules in the field of UV-polarized light, are in a good agreement with our experimental data (solid lines in Figure 2.8):

$$k = 0.01, C = A/W = 0.1 \text{ m}^2/\text{W}, D = 0.00026 \text{ s}^{-1} \quad (2.13)$$

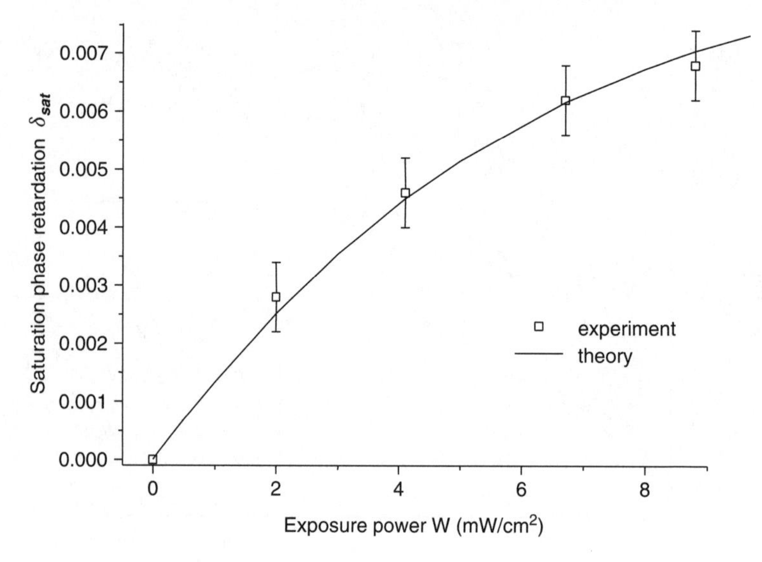

Figure 2.9 Dependence of the relative saturation phase retardation δ_{sat} on power W of the activated UV light [18]. Solid lines indicate the numerical solution of the diffusion model (2.1)–(2.2)

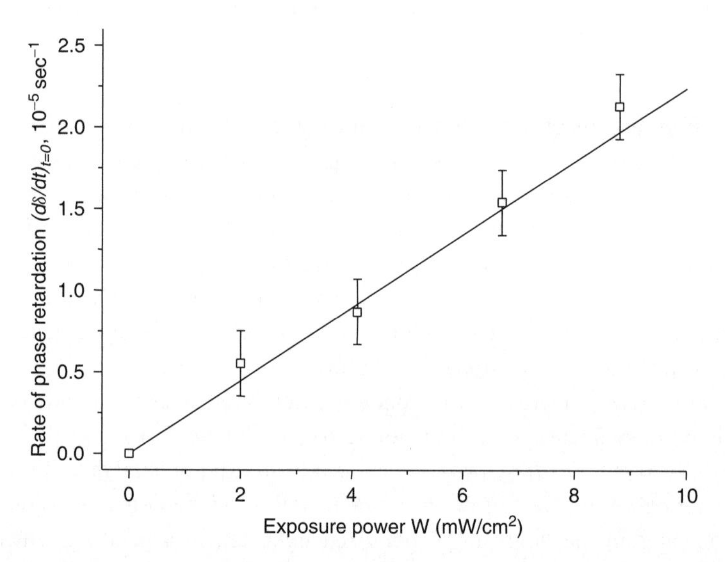

Figure 2.10 Dependence of the initial rate of increase of photoinduced phase retardation $d\delta/dt|_{t=0} = 0$ on power W of the activated UV light [18]. The solid line indicates the numerical solution of the diffusion model (2.1)–(2.2)

Using this data we can determine the relative order parameter s. For example, the relative order parameter according to the saturation level for irradiation power $W =$ 8.8 mW/cm^2 is $s = 0.67$, and for $W = 2.0$ mW/cm^2 it is $s = 0.3$ (Figure 2.9). The saturation value of the phase retardation δ_{sat} shows a nonlinear increase with the power of the activated UV light beam (Figure 2.9). The linear proportionality $s \approx W$ is observed only for sufficiently small values of W (Figure 2.9). However, the exact numerical solution of the diffusion equation (2.1) provides much better agreement (solid line in Figure 2.8).

The measured relaxation dependencies of SD1 azo-dye samples for the same intensity of light, but different exposure times, is shown in Figure 2.11.

The sample was illuminated by linear polarized light for some time (t_{exp}) and after this the UV source was switched off, but the registration of the phase retardation was continued up to the time when changes in the phase retardation δ were not observed. The results for different illumination times are shown in Figure 2.11 [18]. The diffusion process, when the molecular interaction is

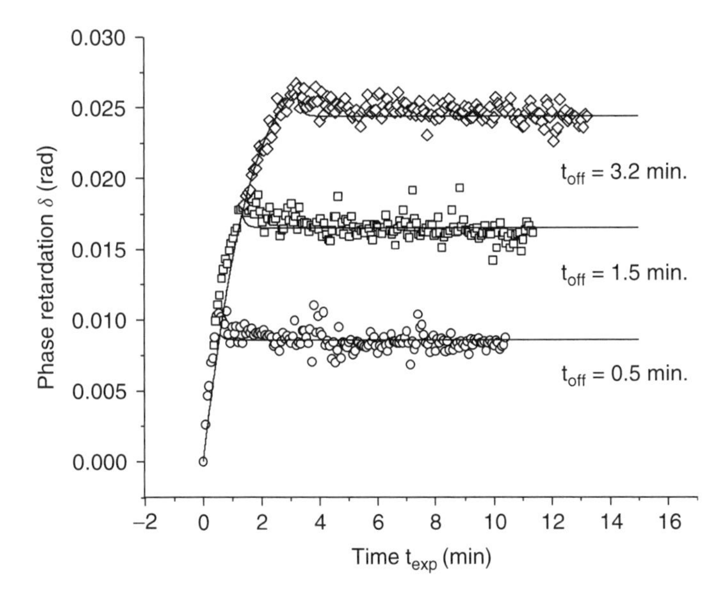

Figure 2.11 The relaxation of azo-dye layer birefringence, proportional to the order parameter after switching off the activated light for different exposure times t_{off} shown in the figure [18]. The modified potential in the diffusion model (2.9)–(2.11) includes the effect of molecular interaction

neglected, cannot explain the order parameter relaxation observed in the experiment (Figure 2.11). Thus the approximation of the 'frozen' potential (2.9)–(2.11) must be used, as the azo-dye molecule is affected by an average molecular field that increases with the exposure time (solid lines in Figure 2.11). It can be seen that for certain values of $A = 0.1$, $a/kT = -5.5$, the approximation of the 'frozen' potential provides a good coincidence between theory and experiment (Figure 2.11).

The proposed diffusion mechanism does not involve any photochemical or structural transformation of azo-dye molecules. Good agreement with experiment was observed even in the case of negligible molecular interaction: that is, (i) the saturation value of birefringence δ_{sat} increases with the intensity of the activated light and is well described by the model (Figure 2.8); (ii) the birefringence increases with the exposure time and reaches the saturation level δ_{sat} (Figure 2.9); and (iii) the initial rate of change of the birefringence $d\delta/dt|_{t=0}$ is proportional to the illumination power (Figure 2.10). A 'frozen' potential in a diffusion equation 'remembers' the final value of the averaged order parameter at the moment when the light is switched off. Thus the effect of an average molecular field is taken into account. Good agreement is found with the observed experimental value. The photoinduced order parameter does not relax to zero and keeps its value when the exposure source is switched off (Figure 2.11).

2.2.2 Polarized Absorption Spectra

In order to study the photo-reorientation of the azo-dye molecules (water-soluble sulfonic dye SD1) by linearly polarized UV exposure, the polarized absorption spectra of the layer before and after irradiation with linearly polarized UV light were measured, using incident light with polarization directions parallel and perpendicular to the polarization direction of the activated linearly polarized UV light [18, 47]. Figure 2.12 shows the polarized absorption spectra (absorbance or optical density) before (curve 1) and after (curves 2 and 3) UV irradiation.

Before irradiation, the absorption of the azo-dye layer does not depend on the polarization of the light used in the measurements. After irradiation by linearly polarized UV light, the absorption of light with a polarization direction parallel to the direction of the activated light ($D_{\|}$) decreases (curve 2, Figure 2.12) while the one with the perpendicular direction ($D_{\perp}$) increases (curve 3, Figure 2.12). The evolution of the polarized absorption spectra after UV illumination does not reveal any noticeable contribution from photochemical reactions [18, 47], as the

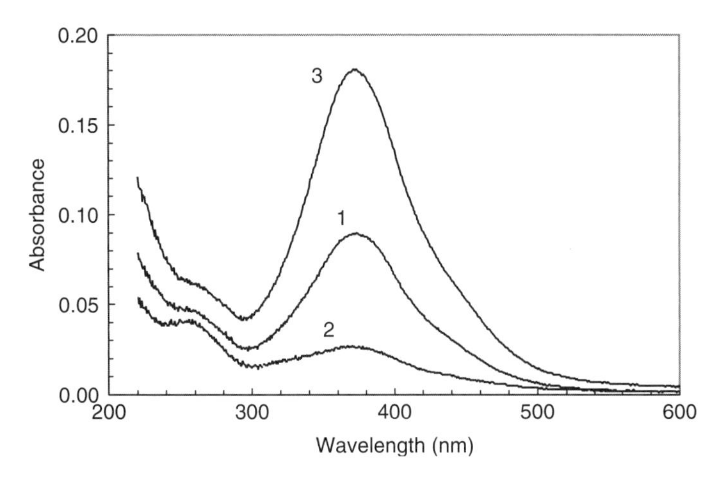

Figure 2.12 Absorption spectra of SD1 layer before polarized UV exposure (curve 1). Curves 2 and 3 show the polarized absorption spectra after exposure to polarized UV light in the direction parallel ($D_{\|}$) and perpendicular ($D_{\perp}$) to the activating light polarization accordingly

average absorption

$$D_{ave} = (D_{\|} + 2D_{\perp})/3 \tag{2.14}$$

remains the same for any fixed value of the exposure time (Figure 2.12). The order parameter S of the azo-dye chromophores can be expressed as [46]

$$S = (D_{\|} - D_{\perp})/(D_{\|} + 2D_{\perp}) \tag{2.15}$$

The order parameter of SD1 is equal to -0.4 at $\lambda_m = 372$ nm (absorption maximum), which is 80% of its maximum absolute value ($S_m = -0.5$ in our case).

2.2.3 Modifications: Repeated Cis-Trans Photoisomerization Reaction Resulting in the Reorientation of the Backbone Structure

The change in molecular orientation of a film of polyamic acid (PAA) with azobenzene units in the backbone structure, induced by irradiation with linearly polarized ultraviolet light (LPUVL irradiation), has been investigated [22, 23]. The PAA backbone structure rotated towards the plane perpendicular to the polarization direction of LPUVL. This angular reorientation occurred through repeated photoisomerization reactions of the azobenzene unit. The change in molecular orientation caused by thermal imidization following LPUVL irradiation has also been studied [23]. Significant enhancement of the molecular order was observed. As a result, polyimide films with an in-plane molecular order much greater than that induced by mechanical rubbing were obtained [22, 23]. The enhancement of the molecular order was tentatively attributed to crystallization of the film caused by thermal imidization. Since the in-plane molecular order of the polyimide film can be controlled over a wide range by varying the LPUVL exposure, the photoinduced alignment method used in this study shows promise as an alternative to the conventional rubbing technique. Thus a thermally stable photoaligning layer based on polyimide structure can be obtained.

2.3 Crosslinking in Cinnamoyl Side-Chain Polymers

The crosslinking of polyvinyl 4-methoxy-cinnamate (PVMC) under the action of linearly polarized light was first observed in [1], where the $(2 + 2)$ cycloaddition reaction of the properly located cinnamoyl fragments was mentioned [10]. However, the first experiment of LC aligning by PVMC film, where the alignment material was clearly identified and the mechanism explained, appeared in 1992 [4, 5]. The results of LC aligning by photo-polymerized PVMC films under the action of LPUVL ($P_{\|}$) light ($\lambda = 320$ nm) are summarized as follows [4] (see Figure 2.13):

1. Linear photo-polymerization (LPP) leads to a preferred depletion of the cinnamic side-chain molecules along $P_{\|}$ due to the $(2 + 2)$ cycloaddition reaction.
2. This causes an anisotropic distribution of cyclobutane molecules with their long axis preferably aligned perpendicular to $P_{\|}$.
3. LC molecules are also aligned perpendicular to $P_{\|}$ (the axis of the preferred alignment of cyclobutane molecules) due to the van der Waals (or dispersion) interaction.

Many publications concerning the mechanism of the photoalignment in various cinnamate side-chain polymers, which provide a homogeneous LC alignment, were reviewed in [8]. According to some of them, the LC alignment is parallel to the polarization direction of the LPUVL. The effect may take place due to both the photochemical *cis-trans* isomerization and dimerization processes.

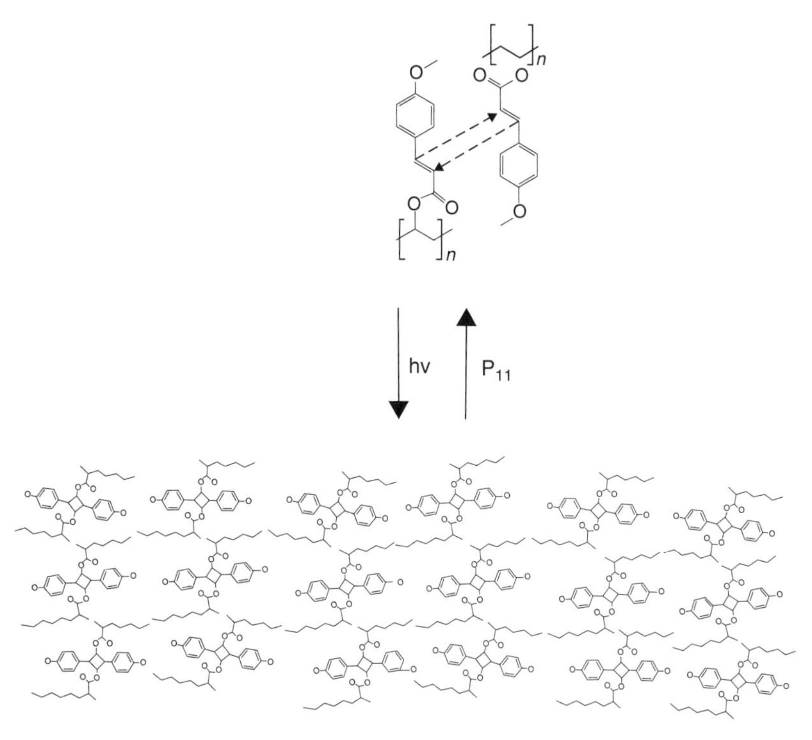

Figure 2.13 The mechanism of crosslinking in polyvinyl 4-methoxy-cinnamate [4]. Reproduced from M. Schadt, K. Schmitt, V. Kozenkov, and V. Chigrinov, Surface-induced parallel alignment of liquid crystals by linearly polymerized photopolymers. *Japanese Journal of Applied Physics* **31**, 2155 (1992), Institute of Pure and Applied Physics

Coumarin side-chain polymers can be also crosslinked by the $(2 + 2)$ cycloaddition reaction, and it is clear that *cis-trans* isomerization cannot occur [8]. The fraction of dimerized product is much greater in this case and LC alignment is parallel to the polarization of the activating UV light [8, 48] (Figure 2.14).

The superior properties of these materials in comparison with cinnamate side-chain polymers were first reported in [49]. The photoalignment orientation of the crosslinking of the polymer LC film with cinnamoyl and biphenyl fragments parallel to the electric field of the incident LPUVL was observed also by Kawatsuki *et al.* [50–52].

The other crosslinking materials suitable for the photoaligning applications were also mentioned elsewhere. In particular, low-molecular-weight composites

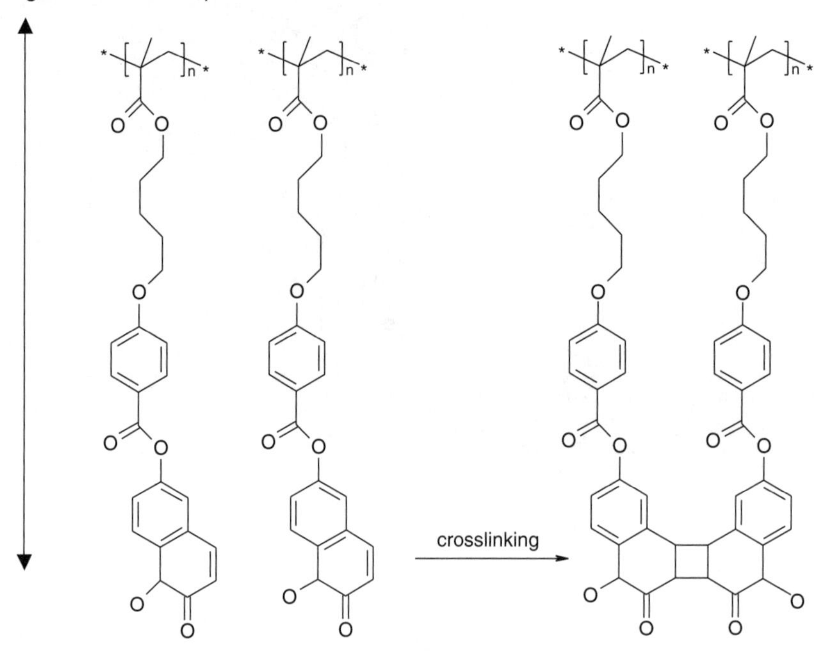

Figure 2.14 Crosslinking in a coumarin side-chain polymer [8]. The LC alignment is parallel to the polarization direction. Reproduced from M. O'Neill and S. M. Kelly, Photoinduced surface alignment for liquid crystal displays. *Journal of Physics D: Applied Physics* **33**, R67 (2000), Institute of Physics

(LMWPCs) [53], polysiloxane cinnamate side-chain polymers [54], and others [55–61]. Certain photoaligned side-chain polymer materials obtained by the crosslinking effect are shown in [8]. The photoalignment properties of poly(vinyl-cinnamate) (PVCN) photopolymer materials were studied in [62, 63], including their fluorinated modification (PVCN-F) [64, 65].

2.4 Photodegradation in Polyimide Materials

As polyimide (PI) materials possess high-temperature stability and were usually used in the LCD industry, it was highly desirable to modify them for photoalignment applications. The first results in this field showed that polarized light at $\lambda = 257$ nm can induce LC alignment perpendicular to the direction of the polarized UV light [6]. Thus the direction of the maximum density of unbroken polyimide chains, which is perpendicular to the light polarization, defines the direction of LC alignment. The direction of the photoalignment can be changed by varying the direction of the UV light polarization [66]. Thus before UV exposure, PI chains in the film are randomly oriented. PI chains parallel to the exposed UV polarization are selectively decomposed by UV exposure, and the corresponding photoproducts become randomly relocated in PI film. The residual PI chains perpendicular to the exposed UV polarization remain unchanged and cause the anisotropic van der Waals forces to align LC molecules along its optical axis [21].

The anisotropy of the PI film surface induced by polarized UV light was investigated using polarized IR absorption of the 1244 cm^{-1} band. It was found that the decomposition rate of the PI chains oriented parallel to the polarization direction of the UV light is greater only by about 23% of that oriented perpendicular to it [67]. UV light with a wavelength of $\lambda = 290$ nm was found to be the most suitable for causing anisotropic decomposition of the polymer chain, as the anisotropy of the absorption of the light polarized parallel and perpendicular to the PI chain direction becomes a maximum [68]. So the larger surface anisotropy can make the photoaligned PI materials as efficient as the usual PI aligning agents, prepared by rubbing. The decomposition process as obtained by spectroscopic analysis involves the weakest bond breaking to form free radicals, which are responsible for the oxidation reaction [69].

There are several drawbacks for LC photoaligning materials prepared by photochemical reaction and by photodegradation in particular:

1. The small value of the order parameter (Equation (2.3)) and the corresponding low value of the induced optical anisotropy and dichroism [22, 67].
2. The order parameter is very sensitive to the exposure time and chemical content of the substance and has to be accurately controlled. For sufficiently large exposure times the order parameter goes through a maximum to zero (Figures 2.15 and 2.16) [20, 62, 70]. The latter was first mentioned in the early work of Kozenkov *et al.* [71], who predicted this effect for photosensitive layers, where molecular movement is hindered. Indeed, for sufficiently high exposure time the photoinduced transformation will take place not only for the molecular units, with which the absorption oscillator forms a small angle with

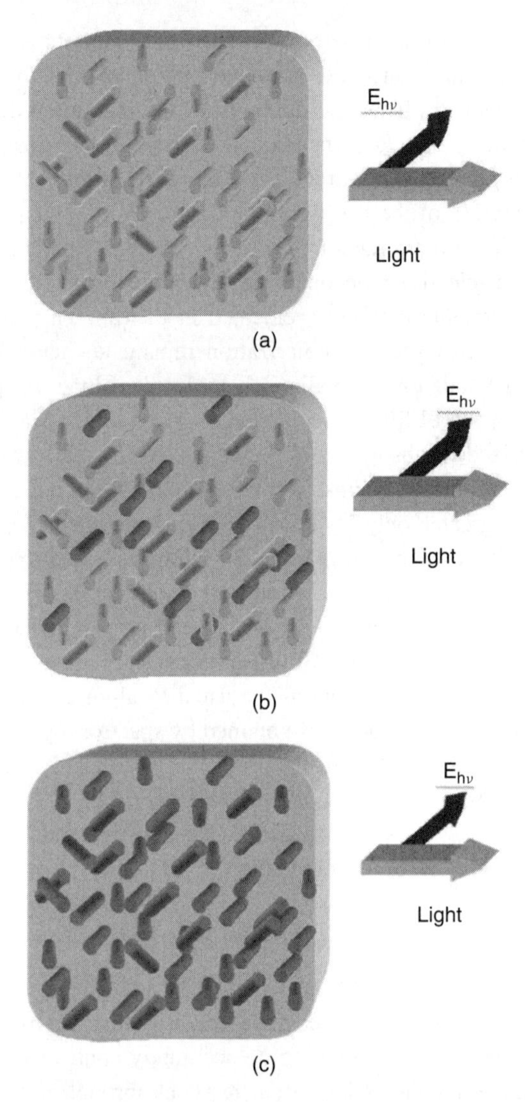

Figure 2.15 Photochemical reaction in the photosensitive layer where the molecular motion is hindered. After a sufficiently high exposure time the polarization of the activated light $E_{h\nu}$ will induce photochemical transformation not only in the molecular units having the absorption oscillator parallel to the polarization, but also within the whole molecular ensemble of the photosensitive layer [10, 71]

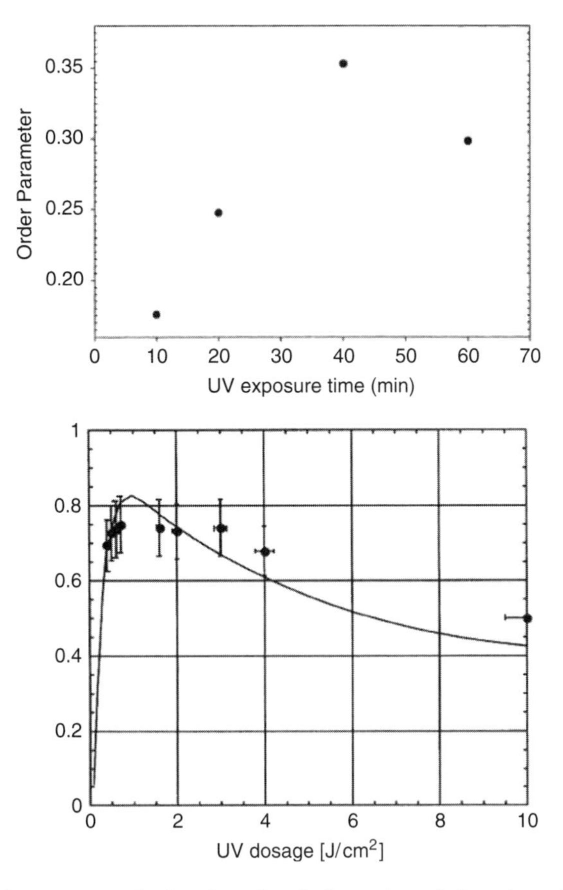

Figure 2.16 Order parameter in the photochemical reaction of photodegradation as a function of UV exposure time (upper, [70]) and UV exposure dosage (lower, [20]). Reproduced from M. Hasegawa, Modeling of photoinduced optical anisotropy and anchoring energy of polyimide exposed to linearly polarized deep UV light. Japanese Journal of Applied Physics **38**, L457 (1999), Institute of Pure and Applied Physics

the activating light polarization, but also for the entire molecular ensemble, thus leading to a zero order (Figure 2.15).

3. Contamination of the initial substance by the by-products of the photodegradation is possible in certain cases. The by-products can produce ions and initiate the image sticking effect and flicker [72], as well as reduce the thermal stability of the LC alignment [73].

4. Low sensitivity in PI decomposition. The reported sensitivity for photodegradable polyimide materials was 3–12.7 J/cm^2 for aromatic PIs, when broadband UV light was used [74].

Certain research has been conducted to improve the photoaligning performance of PI layers. A novel method for LC alignment using *in situ* exposure of the PI film during imidization resulted in a sufficient increase of the thermal stability in comparison with the conventional method, which employed UV exposure after imidization [75]. However, the thermal stability is considerably deteriorated if the exposure time becomes too high. New materials based on PI have appeared, where the main mechanism of the LC photoalignment was not photodegradation, but reorientation of the main chain of the polymer perpendicular to the polarization of the light, as shown in [22, 23, 76, 77]. Laser light with a wavelength of $\lambda = 355$ nm was used to activate the PI film and no obvious structural change was observed in this case. New crosslinking and solvent-soluble PIs, containing cinnamate side chains, were also developed for LC aligning [78]. The LC alignment was obtained perpendicular to the direction of UV light ($\lambda = 350$ nm) polarization and remained intact after being heated at 85 °C for 450 h. New PIs with pendant cinnamate groups were also suitable for photoaligning [79]. However, the best thermal stability of the photoaligned layer was observed at the lowest cinnamate content.

To explain the phenomenon of decomposition, Hasegawa applied the reaction kinetic model [20]. The peak value of the order parameter of the LC mixture attached to the polyimide surface was achieved at an exposure energy of about 1 J/cm^2 and fitted the experimental data well (Figure 2.16).

The simple, bare solid substrates that can adsorb organic molecules and are sufficiently transparent to UV light can be photo-oriented by this light [24]. This can be naturally interpreted in terms of a selective desorption. A molecule that absorbs a UV photon leaves the surface, and such an evaporation is governed by the anisotropic light absorption rate. This rate is a maximum for molecules with their dipoles parallel to the light polarization plane and a minimum for the normal direction. As a result, the relative excess molecules of the latter orientation form grooves along this normal direction. This is analogous to a kind of light rubbing in a direction normal to the light polarization plane: the rubbing readily removes the molecules lying across the rubbing direction, whereas those lying along this direction have less chance to get scratched away. Light with a rotating polarization cleans the surface of molecules with any orientation, creating no anisotropy in accord with the experiment [24]. The result points to the possibility that the search for photoaligning surfaces can continue not only by studying complicated combinations of substrate materials and coating, but also by a much simpler study of the light rubbing of transparent bare solid substrates with adsorbed molecules.

2.5 Photoinduced Order in Langmuir–Blodgett Films

All the mechanisms described above can be considered to describe the photoinduced order in Langmuir–Blodgett (LB) films. Photoinduced optical anisotropy as a result of azo-dye reorientation in LB films has already been mentioned above (Section 2.2) [17, 44, 45]. Actually, the photoinduced molecular reorientation was not associated with photochemical reactions or *cis-trans* isomerization [80]. In this case the azo-dye aligning mechanism was very similar to the *diffusion rotation* of the molecular units described above (Section 2.2.1). A homogeneous alignment with excellent optical quality and acceptable storage stability was generated by illuminating an LB monolayer of a PI-bearing p-cyanoazobenzene, confirming that the in-plane LC alignment regulation was realized by an azobenzene *command monolayer* (Section 2.1.1) [81]. The *crosslinking* can also be used to form a photoalignment order in LB films [82].

References

[1] E. D. Kvasnikov, V. M. Kozenkov, and V. A. Barachevskii, Birefringence in polyvinyl cinnamate films, induced by polarized light. *Dokladi Akademii Nauk SSSR* **237**, 633 (1977) (in Russian).

[2] K. Ichimura, Y. Suzuki, T. Seki, A. Hosoki, and K. Aoki, Reversible change in alignment mode of nematic liquid crystals regulated photochemically by 'command surfaces' modified with an azobenzene monolayer. *Langmuir* **4**, 1214 (1988).

[3] W. M. Gibbons, P. J. Shannon, S. T. Sun, and B. J. Swetlin, Surface mediated alignment of nematic liquid crystals with polarized laser light. *Nature* **351**, 49 (1991).

[4] M. Schadt, K. Schmitt, V. Kozenkov, and V. Chigrinov, Surface-induced parallel alignment of liquid crystals by linearly polymerized photopolymers. *Japanese Journal of Applied Physics* **31**, 2155 (1992).

[5] A. G. Dyadyusha, T. Ya. Marusii, Yu. A. Reznikov, A. I. Khizhnyak, and V. Yu. Reshetnyak, Orientational effect due to a change in the anisotropy of the interaction between a liquid crystal and a bounding surface. *JETP Letters* **56**, 17 (1992).

[6] M. Hasegawa and Y. Taira, Nematic homogeneous photoalignment by polyimide exposure to linearly polarized UV. *Journal of Photopolymer Science and Technology* **8**, 241 (1995).

[7] L. M. Blinov, Photoinduced molecular reorientation in polymers, Langmuir Blodgett films and liquid crystals. *Journal of Nonlinear Optical Physics and Materials* **5**, 165 (1996).

[8] M. O'Neill and S. M. Kelly, Photoinduced surface alignment for liquid crystal displays. *Journal of Physics D: Applied Physics* **33**, R67 (2000).

[9] M. Schadt, Liquid crystal displays and novel optical thin films enabled by photoalignment. *Molecular Crystals and Liquid Crystals* **364**, 151 (2001).

[10] V. G. Chigrinov, V. M. Kozenkov, and H.-S. Kwok, New developments in photoaligning and photo-patterning technologies: physics and applications. In *Optical Applications of Liquid Crystals*, Ed. L. Vicari, Institute of Physics Publishing, Bristol and Philadelphia, 201–244 (2003).

[11] V. P. Shibaev, S. A. Kostromin, and S. A. Ivanov, *Polymers as Electrooptical and Photooptical Active Media*, Springer-Verlag, Berlin, 37–110 (1996).

[12] T. G. Pedersen, P. S. Ramanujam, P. M. Johansen, and S. Hvilsted, Quantum theory and experimental studies of absorption spectra and photoisomerization of azobenzene polymers. *Journal of the Optical Society of America* **15**, 2721 (1998).

[13] T. G. Pedersen, P. M. Johansen, N. C. R. Holme, and P. S. Ramanujam, Theoretical model of photoinduced anisotropy in liquid crystalline azobenzene side-chain polyesters. *Journal of the Optical Society of America* **15**, 1120 (1998).

[14] T. G. Pedersen and P. M. Johansen, Mean-field theory of photoinduced molecular reorientation in azobenzene liquid crystalline side-chain polymers. *Physical Review Letters* **79**, 2470 (1997).

[15] Y. B. Gaididei, P. L. Christiansen, and P. S. Ramanujam, Theory of photoinduced deformation of molecular films. *Applied Physics B*, **74**, 139 (2002).

[16] O. V. Yaroshchuk, A. D. Kiselev, Yu. Zakrevskyy, T. Bidna, J. Kelly, L.-C. Chien, and J. Lindau, Photo-induced three dimensional orientational order in side chain liquid crystalline azopolymers. *Physical Review E*, **68**, 011803 (2003).

[17] S. P. Palto, N. M. Shtykov, V. A. Khavrichev, and S. G. Yudin, Photoinduced optical anisotropy in Langmuir–Blodgett films. *Molecular Materials* **1**, 3 (1992).

[18] V. Chigrinov, S. Pikin, A. Verevochnikov, V. Kozenkov, M. Khazimullin, J. Ho, D. Huang, and H.-S. Kwok, Diffusion model of photoaligning in azo-dye layers. *Physical Review E* **69**, 061713 (2004).

[19] J. Chen, D. L. Johnson, P. J. Bos, X. Wang, and J. L. West, Model of liquid crystal alignment by exposure to linearly polarized ultraviolet light. *Physical Review E* **54**, 1599 (1996).

[20] M. Hasegawa, Modeling of photoinduced optical anisotropy and anchoring energy of polyimide exposed to linearly polarized deep UV light. *Japanese Journal of Applied Physics* **38**, L457 (1999).

[21] M. Nishikawa, B. Taheri, and J. L. West, Mechanism of unidirectional liquid-crystal alignment on polyimides with linearly polarized ultraviolet light exposure. *Applied Physics Letters* **72**, 2403 (1998).

[22] B. Park, Y. Jung, H. Choi, H. Hwang, Y. Kim, S. Lee, S. Jang, M. Kakimoto, and H. Takezoe, Thermal and optical stabilities of photoizomerazible polyimide layers for nematic liquid crystal alignment. *Japanese Journal of Applied Physics* **37**, 5663 (1998).

[23] K. Sakamoto, K. Usami, M. Kikegawa, and S. Ushioda, Alignment of polyamic acid molecules containing azobenzene in the backbone structure: effects of polarized ultraviolet light irradiation and subsequent thermal imidization. *Journal of Applied Physics* **93**, 1039 (2003).

[24] V. G. Nazarenko, O. P. Boiko, A. B. Nych, Yu. A. Nastishin, V. M. Pergamenshchik, and P. Bos, Selective light-induced desorption: the mechanism of photoalignment of liquid crystals at adsorbing solid surfaces. *Europhysics Letters* **75**, 448 (2006).

[25] M. Schonhoff, M. Mertesdorf, and M. Losche, Mechanism of photoreorientation of azobenzene dyes in molecular films. *Journal of Physical Chemistry* **100**, 7558 (1996).

[26] K. Ichimura, Y. Hayashi, H. Akiyama, T. Ikeda, and N. Ishizuki, Photo-optical liquid crystal cells driven by molecular rotors. *Applied Physics Letters* **63**, 449 (1993).

[27] K. Aoki, Y. Kawanishi, T. Seki, M. Sakuragi, T. Tamaki, and K. Ichimura, Reversible alignment change of liquid crystals induced by photochromic molecular films: properties of azobenzene chromophores covalently attached to silica surfaces. *Liquid Crystals* **19**, 119 (1995).

[28] H. Knobloch, H. Orendi, M. Buchel, T. Seki, Sh. Ito, and W. Knoll, Command surface controlled liquid crystalline waveguide structures as optical information storage. *Journal of Applied Physics* **76**, 8212 (1994).

[29] L. Gui, P. Xie, R. Zhang, and T. Yang, Photo-driven liquid crystal cell with high sensitivity. *Liquid Crystals* **26**, 1541 (1999).

[30] L. Komitov, C. Ruslim, Y. Matsuzawa, and K. Ichimura, Photoinduced anchoring transitions in a nematic doped with azo dyes. *Liquid Crystals* **27**, 1011 (2000).

[31] Y. Liao, Ch. Hsu, and S. T. Wu, Dipping and photo-induced liquid crystal alignment using silane surfactants. *Japanese Journal of Applied Physics* **39**, L-90 (2000).

[32] K. Ichimura, H. Tomita, and K. Kudo, Command surfaces 14 [1]. Photoregulation of in-plane alignment of a liquid crystal by the photoisomerization of stilbenes chemisorbed on a substrate silica surface. *Liquid Crystals* **20**, 161 (1996).

[33] C. Jones and S. Day, Shedding light on alignment. *Nature* **351**, 15 (1991).

[34] R. Rosenhauer, Th. Fischer, S. Czapla, J. Stumpe, A. Vinuales, M. Pinol, and J. L. Serrano, Photo-induced alignment of LC polymers by photoorientation and thermotropic self-organization. *Molecular Crystals and Liquid Crystals* **364**, 295 (2001).

[35] A. G. Tereshchenko, L. I. Shanskii, O. V. Yaroshchuk, and Yu. Lindau, Photoinduced optical anisotropy in azopolymer films. *Optics and Spectroscopy* **83**, 808 (1997).

[36] O. Yaroshchuk, T. Bidna, M. Dumont, and J. Lindau, Photoinduced 3D orientational order in liquid crystalline azopolymers studied by the method of attenuated total reflection. *Molecular Crystals and Liquid Crystals* **409**, 229 (2204).

[37] O. Yaroshchuk, Yu. Zakrevskyy, S. Kumar, J. Kelly, L.-C. Chien, and J. Lindau, Three dimensional order in the bulk and on the surface of the polymer films and its effect on liquid-crystal alignment. *Physical Review E* **69**, 011702 (2004).

[38] O. Yaroshchuk, M. Dumont, Yu. Zakrevskyy, T. Bidna, and J. Lindau, Molecular structure of azopolymers and photoinduced 3D orientational order. 1. Azobenzene polyesters. *Journal of Physical Chemistry B* **108**, 4647 (2004).

[39] S. T. Sun, W. M. Gibbons, and P. J. Shannon, Alignment of guest-host liquid crystals with polarized laser light. *Liquid Crystals* **12**, 869 (1992).

[40] P. J. Shannon, W. M. Gibbons, and S. T. Sun, Patterned optical properties in photopolymerized surface-aligned liquid-crystal films. *Letters to Nature* **368**, 532 (1994).

[41] W. M. Gibbons, T. Kosa, P. Palffy-Muhoray, P. J. Shannon, and S. T. Sun, Continuous grey-scale image storage using optically aligned nematic liquid crystals. *Letters to Nature* **377**, 43 (1995).

[42] B. Umanskii, N. Novoseletskii, S. Torgova, and G. Dorozhkina, Azo-dyes as photoalignment materials for liquid crystals. *Molecular Crystals and Liquid Crystals* **412**, 313 (2004).

[43] J. West, L. Su, and Yu. Reznikov, Photoalignment using adsorbed dichroic molecules. *Molecular Crystals and Liquid Crystals* **412**, 313 (2004).

[44] V. M. Kozenkov, S. G. Yudin, E. G. Katyshev, S. P. Palto, V. T. Lazareva, and V. A. Barachevskiy, Photoinduced optical anisotropy in multilayer Langmuir films. *Soviet Technical Physics Letters* **12**, 525 (1986).

[45] M. I. Barnik, V. M. Kozenkov, N. M. Shtykov, S. P. Palto, and S. G. Yudin, Photoinduced optical anisotropy in Langmuir-Blodgett films. *Journal of Molecular Electronics* **5**, 53 (1989).

[46] V. G. Chigrinov, *Liquid Crystal Devices: Physics and Applications*, Artech House, London (1999).

[47] V. Chigrinov, E. Prudnikova, V. Kozenkov, H. Kwok, H. Akiyama, H. Kawara, H. Takada, and H. Takatsu, Synthesis and properties of azo dye aligning layers for liquid crystal cells. *Liquid Crystals* **29**, 1321 (2002).

[48] N. Kawatsuki, K. Goto, and T. Yamamoto, Photoinduced anisotropy and photoalignment of nematic liquid crystals by a novel polymer liquid crystal with a coumarin-containing side group. *Liquid Crystals* **28**, 1171 (2001).

[49] M. Schadt, H. Seiberle, and A. Schuster, Optical patterning of multidomain liquid-crystal displays with wide viewing angles. *Nature* **381**, 212 (1996).

[50] N. Kawatsuki, H. Tatsuka, T. Yamamoto, and H. Ono, Photoregulated liquid crystal alignment on photoreactive side-chain liquid crystalline polymer. *Japanese Journal of Applied Physics* **36**, 6464 (1997).

[51] N. Kawatsuki, T. Yamamoto, and H. Ono, Photoinduced alignment control of photoreactive side-chain polymer liquid crystal by linearly polarized ultraviolet light. *Applied Physics Letters* **74**, 935 (1999).

[52] N. Kawatsuki, H. Takatsuka, and T. Yamamoto, Thermally stable photoalignment layer of a novel photo-crosslinkable polymethacrylate for liquid crystal display. *Japanese Journal of Applied Physics* **40**, L209 (2001).

[53] O. Yaroshchuk, L. G. Cada, M. Sonpatki, and L.-C. Chien, Liquid-crystal photoalignment using low-molecular-weight photo-cross-linkable composites. *Applied Physics Letters* **79**, 30 (2001).

[54] O. Yaroshchuk, G. Pelzl, G. Pirwitz, Yu. Reznikov, H. Zaschke, J.-H. Kim, and S. B. Kwon, Photosensitive materials on a base of polysiloxane for the alignment of nematic liquid crystals. *Japanese Journal of Applied Physics* **36**, 5693 (1997).
[55] M. Kimura, Sh. Nakata, Y. Makita, Y. Matsuki, A. Kumano, Y. Takeuchi, and H. Yokoyama, Strong liquid crystal anchoring on photoalignment copolymer films containing ω(4-chalconyloxy)alkyl side groups. *Japanese Journal of Applied Physics* **40**, L352 (2001).
[56] S. Perny, P. Le Barny, J. Delaire, T. Buffeteau, and C. Sourisseau, Molecular orientation and liquid crystal alignment properties of new cinnamate-based photocrosslinkable polymers. *Liquid Crystals* **27**, 341 (2000).
[57] H. Murai, T. Nakata, and H. Goto, Liquid crystal photoalignment layers made from aromatic bismaleimides. *Liquid Crystals* **5**, 669 (2002).
[58] J.-Y. Kim, J.-Y. Hwang, T.-H. Kim, D.-S. Seo, and J.-U. Lee, Liquid crystal alignment effect on photopolymer surface using an acrylate unit photopolymerized by a photoinitiator. *Japanese Journal of Applied Physics* **42**, L54 (2003).
[59] J.-Y. Hwang, D.-S. Seo, J.-Y. Kim, and T.-H. Kim, Aligning capabilities of nematic liquid crystal on photopolymer-based N-(phenyl)maleimide hopolymers. *Japanese Journal of Applied Physics* **42**, 194 (2003).
[60] T. Mihara, M. Tsutsumi, and N. Koide, Preparation of photo-crosslinked films of new chalcone-based side chain type liquid crystalline polymers and its application to alignment film. *Molecular Crystals and Liquid Crystals* **412**, 247 (2004).
[61] J.-Y. Hwang, D.-S. Seo, and D.-H. Suh, Liquid crystal aligning capabilities and EO characteristics of the photoaligned TN-LCD on a photo-crosslinkable polyamides based polymer. *Molecular Crystals and Liquid Crystals* **412**, 247 (2004).
[62] G. P. Bryan Brown and I. C. Sage, Photoinduced ordering and alignment properties of polyvinylcinnamates. *Liquid Crystals* **20**, 825 (1996).
[63] K. Rajesh, S. Masuda, R. Yamaguchi, and S. Sato, Alignment of liquid crystal on poly (vinyl cinnamate) photopolymer and anchoring direction. *Japanese Journal of Applied Physics* **36**, 4404 (1997)
[64] D. Andrienko, A. Dyadyusha, Y. Kurioz, Y. Reznikov, F. Barbet, D. Bormann, M. Warenghem, and B. Khelifa, Photoalignment of pentyl-cyanobiphenyl on the fluorinated polyvinyl-cinnamates induced by UV and visible light. *Molecular Crystals and Liquid Crystals* **329**, 219 (1999).
[65] I. Gerus, A. Glushchenko, S. B. Kwon, V. Reshetnyak, and Yu. Reznikov, Anchoring of a liquid crystal on a photoaligning with varying surface morphology. *Liquid Crystals* **28**, 1709 (2001).
[66] J. L. West, X. Wang, Y. Li, and J. R. Kelly, Polarized UV-exposed polyimide films for liquid-crystal alignment. *SID'95 Digest*, p. 703 (1995).
[67] K. Sakamoto, K. Usami, M. Watanabe, R. Arafune, and S. Ushioda, Surface anisotropy of polyimide film irradiated with linearly polarized ultraviolet light. *Applied Physics Letters* **72**, 1832 (1998).
[68] K. Sakamoto, K. Usami, R. Arafune, and S. Ushioda, Dichroism of polyimide chain for ultraviolet light. *Molecular Crystals and Liquid Crystals* **329**, 393 (1999).

[69] S. Gong, J. Kanicki, L. Ma, and J. Zhong, Ultraviolet-light induced liquid-crystal alignment on polyimide films. *Japanese Journal of Applied Physics* **38**, 5996 (1999).
[70] S.-J. Sung, H.-T. Kim, J.-W. Lee, and J.-K. Park, Photo-induced liquid crystal alignment on polyimide containing fluorine group. *Synthetic Metals* **117**, 277 (2001).
[71] V. M. Kozenkov and V. A. Barachevskii, *Organic Photoanisotropic Materials and Their Application*, Ed. V. A. Barachevsky, Nauka, Leningrad, 89 (1987) (in Russian).
[72] K. H. Yang, K. Tajima, A. Takenaka, and H. Takano, Charge trapping properties of UV-exposed polyimide films for the alignment of liquid crystals. *Japanese Journal of Applied Physics* **35**, L561 (1996).
[73] Y. Wang, C. Xu, A. Kanazawa, T. Shiono, T. Ikeda, Y. Matsuki, and Y. Takeuchi, Thermal stability of alignment of nematic liquid crystal induced by polyimides exposed to linearly polarized light. *Liquid Crystals* **28**, 473 (2001).
[74] Z.-X. Zhong, X. D. Li, S. H. Lee, and M.-H. Lee, Liquid crystal photoalignment material based on chloromethylated polyimide. *Applied Physics Letters* **85**, 2520 (2004).
[75] J.-H. Kim, B. R. Acharya, D. M. Agra, and S. Kumar, Thermal stability of liquid crystal alignment layers prepared by in-situ ultra-violet exposure during imidization of polyimide. *Japanese Journal of Applied Physics* **40**, 2381 (2001).
[76] X.-D. Li, Z.-H. Zhong, S.-H. Lee, G. Ghang, and M.-H. Lee, Liquid crystal photoalignment using soluble photosensitive polyimide. *Japanese Journal of Applied Physics* **45**, 906 (2006).
[77] K. Sakamoto, K. Usami, T. Kanayama, M. Kikegawa, and S. Ushioda, Photoinduced inclination of polyimide molecules containing azobenzene in the backbone structure. *Journal of Applied Physics* **94**, 2302 (2003).
[78] W.-C. Lee, C.-S. Hsu, and S.-T. Wu, Liquid crystal alignment with a photo-crosslinkable and solvent soluble polyimide film. *Japanese Journal of Applied Physics* **39**, L471 (2000).
[79] H.-T. Kim, J.-W. Lee, S.-J. Sung, and J.-K. Park, Synthesis, photoreaction and photo-induced liquid crystals alignment on soluble polyimide with pendant cinnamate group. *Liquid Crystals* **27**, 1343 (2000).
[80] S. Palto, J. Malthete, C. Germain, and G. Durand, On the nature of photoinduced optical anisotropy in diacetylene Langmuir-Blodgett films. *Molecular Crystals and Liquid Crystals* **282**, 451 (1996).
[81] H. Akiyama, K. Kudo, and K. Ichimura, Azimuthal photoregulation of a liquid crystal with an azobenzene-modified polyimide Langmuir–Blodgett monolayer. *Langmuir* **11**, 1033 (1995).
[82] Z. Peng and L. Xuan, Alignment of liquid crystals induced by a photopolymerized self-assembled film. *Liquid Crystals* **32**, 239 (2005).

3

LC-Surface Interaction in a Photoaligned Cell

3.1 Pretilt Angle Generation in Photoaligning Materials

The pretilt angle of LC molecules is one of the most important properties of an alignment layer. Pretilt angles are needed to give a preferred tilting direction for the molecules under an applied voltage. In most of the electro-optical modes, the pretilt angle should be sufficiently high to ensure good performance of the LCD. For instance, in a supertwisted nematic liquid crystal display (STN-LCD) the value of the pretilt angle at the substrates should be high enough to provide perfect switching between the supertwisted and almost homeotropic configurations to avoid hysteresis phenomena [1]. Normal rubbed polyimide alignment layers can provide sufficient pretilt of a few degrees. It is important that photoalignment can also give the same values of the pretilt angle in LC cells.

It is easy to see that the pretilt angle is not defined when normally incident polarized light is used in photoalignment. The two molecular symmetrical pretilt

Photoalignment of Liquid Crystalline Materials: Physics and Applications
V. Chigrinov, V. Kozenkov and H.-S. Kwok

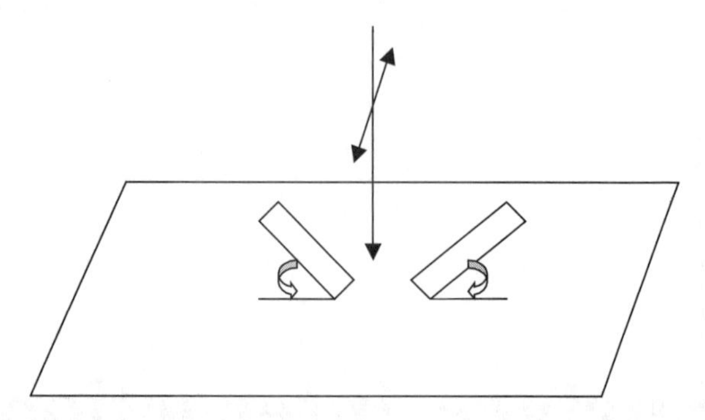

Figure 3.1 The pretilt angle is not defined when normally incident polarized light activates the molecules in the photosensitive layer. The two molecular symmetrical pretilt angles shown here are both possible, because the absorption oscillators are perpendicular to the activated light polarization in both cases. The molecules, similar to those described in Figure 2.4, are shown by rods

angles shown in Figure 3.1 are both possible, because the absorption oscillators are perpendicular to the activated light polarization in both cases.

Thus to induce the pretilt angle, the symmetry of the case should be broken. One possibility to remove the degeneracy of the pretilt alignment direction is to involve the LC flow when the cell is filled with LC [2]. However, even though a pretilt angle is generated, subsequent heating and cooling of the LC cell will again restore the degeneration in the pretilt angle, as the filling direction will be 'forgotten'. Thus this is not a viable solution. A much better way to break the symmetry is simply by changing the incident angle of the light during photoalignment.

There are two ways that this oblique incident method can be carried out. The method of double polarized UV exposure with different polarization and irradiation angles was first proposed by Iimura *et al.* [3] (Figure 3.2).

Normally incident polarized light first changes the random direction of the molecules to an alignment perpendicular to the polarization, i.e. in one plane, and after this, obliquely incident light with a p-polarization chooses an alignment direction perpendicular to the polarization or parallel to the propagation direction (Figure 3.2). However the pretilt angles obtained by Iimura *et al.* were rather small and did not exceed 0.3° [3]. The effect can be partially explained by the fact that the observed LC pretilt angle is dependent not only on the direction of the anisotropic order in the photoaligning layer, but also on the properties of the LC mixture in contact with this layer. In other words, the final result will be dependent

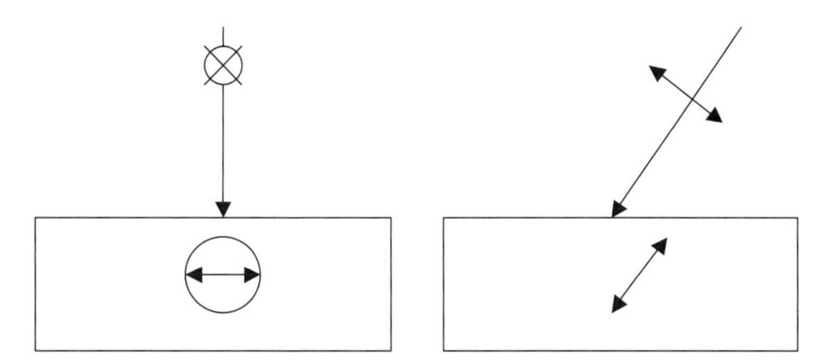

Figure 3.2 Two-step exposure by polarized light to induce the pretilt angle in the photoaligned film [3]. After the first exposure the molecules are arranged in one plane perpendicular to the light polarization. The second exposure chooses the alignment perpendicular to the direction of p-polarization of the obliquely incident light (parallel to the light propagation). Reproduced from Y. Iimura, T. Saitoh, S. Kobayashi and T. Hashimoto, Liquid crystal alignment on photopolymer surfaces exposed by linearly polarized UV light. *Journal of Photopolymer Science and Technology* **8**, 258 (1995), Tokai University

on the interaction of the two order parameters: the photoalignment layer and LC layer [1].

The pretilt angles can also be obtained in a one-step exposure by obliquely incident polarized light, in case the alignment is induced parallel to the polarization direction [4]. Any desired LC pretilt angle from 0° to 90° has been demonstrated, using polyimide films with different molar fractions of fluorine fragments (Figure 3.3).

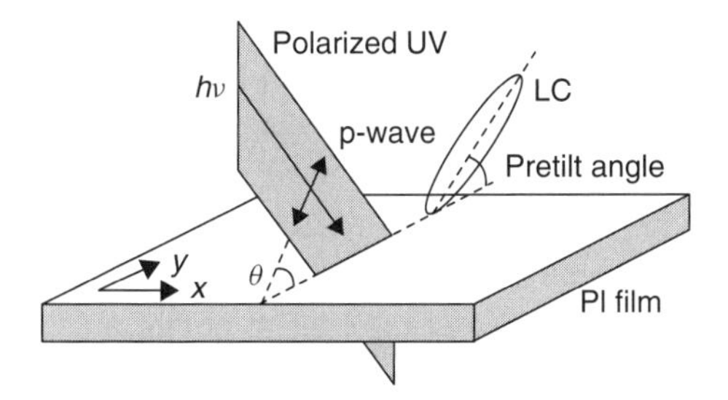

Figure 3.3 Generation of the pretilt angle by a one-step exposure [4]. Reproduced from M. Nishikawa and J. West, Mechanism of generation of pretilt angles on polyimides with a single exposure of polarized ultraviolet light. *Japanese Journal of Applied Physics* **38**, 5183 (1999), Institute of Physics

However, the value of the pretilt angle decreases for sufficiently high exposure times, which is a consequence of the photochemical reaction in polyimide (PI) as discussed earlier (Figure 2.15). Pretilt angles can be induced by a one-step exposure of polarized light, when coumarin-substituted cinnamate derivatives are used (Figure 2.14) [5]. Pretilt angles of about 2° to 6° were produced when the PI film was illuminated by non-polarized obliquely incident UV light [6–8].

The generation of pretilt angles in a nematic liquid crystal (NLC) has been investigated [9] using an *in situ* photoalignment method with polarized UV exposure during imidization of PI on two kinds of PI surfaces with side chains. The pretilt angles of NLC generated using the *in situ* photoalignment method were smaller than those generated using a conventional photoalignment method for a short UV exposure time and did not exceed 2°.

The pretilt angle of LC molecules induced by photoaligned films of PI containing azobenzene in the backbone structure (Azo-PI) was studied in [10]. The photoalignment treatment was single oblique-angle irradiation with non-polarized light at an incidence angle of 45°. It was performed on films of polyamic acid (Azo-PAA) using a light source of wavelength 340–500 nm. The photo-treated films of Azo-PAA were converted into Azo-PI by thermal imidization. The pretilt angle increased with UV exposure, reaching $\sim 3^\circ$ at 880 J/cm^2.

Certain research was undertaken to generate pretilt angles combining the homogeneous alignment produced by UV illuminated PI films doped with surfactants such as lecithine [11]. However, only those angles close to homogeneous alignment (1.8°) or homeotropic alignment (89.5°) can be produced depending on the concentration of lecithine.

Four kinds of exposure method were tested to produce the pretilt angle, using azo-dye materials [12] (Figure 3.4).

To measure the pretilt angle, the LC cells were assembled with substrates antiparallel to the UV irradiation directions and were filled with LC mixture. The crystal rotation method was used to measure their pretilt angles [1]. Pretilt angles of $0.7^\circ, 0.1^\circ, 5.3^\circ$, and 3.8° are obtained with the exposure method (a), (b), (c), and (d), respectively (Figure 3.4). Oblique irradiation with non-polarized light proved to be the most efficient in order to obtain larger pretilt angles. The dependence of the pretilt angle on the exposure energy in method (c) is shown in Figure 3.5. The pretilt angle was temperature stable and did not change after heating the sample at 100 °C for 10 min.

A normal incidence single-exposure system to obtain stable pretilted alignment of NLCs on photoalignment polymer layers was proposed using a photorubbing

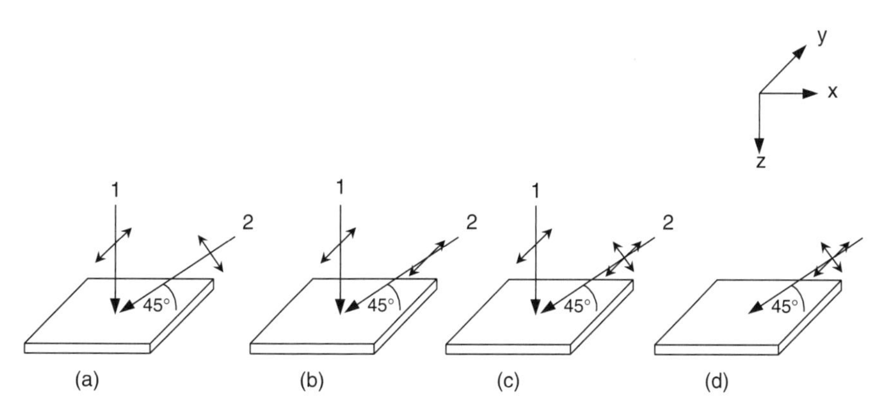

Figure 3.4 Various methods of UV light irradiation to attain the pretilt angle of LC layer on the photoaligned film. Case (a): two-step exposure by y-polarized light vertical to the plane of the azo-dye film (xy plane) and oblique p-polarized light. Case (b): case (a) with s-polarized light used at the second step. Case (c): case (a) with non-polarized light at the second step. Case (d): one-step exposure with oblique non-polarized light. The activated UV light propagates in the xz plane at an oblique angle with the substrate $\theta = 45^\circ$ [12]. Reproduced from V. Chigrinov, E. Prudnikova, V. Kozenkov, H. Kwok, H. Akiyama, H. Kawara, H. Takada, and H. Takatsu, Synthesis and properties of azo dye aligning layers for liquid crystal cells. *Liquid Crystals* **29**, 1321 (2002), IEEE

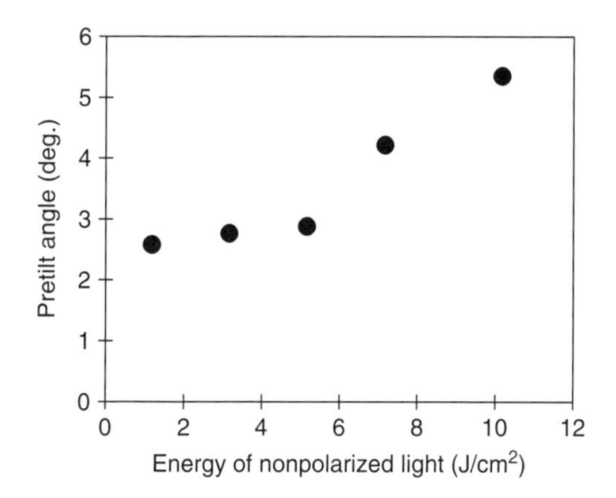

Figure 3.5 Dependence of the pretilt angle on the energy density of the obliquely irradiated non-polarized light after the irradiation of vertically polarized light (Figure 3.4(c)) with an energy density of 10 J/cm^2 [12]. Reproduced from V. Chigrinov, E. Prudnikova V. Kozenkov, H. Kwok, H.Akiyama, H. Kawara, H. Takada, and H. Takatsu, Synthesis and properties of azo dye aligning layers for liquid crystal cells. *Liquid Crystals* **29**, 1321 (2002), IEEE

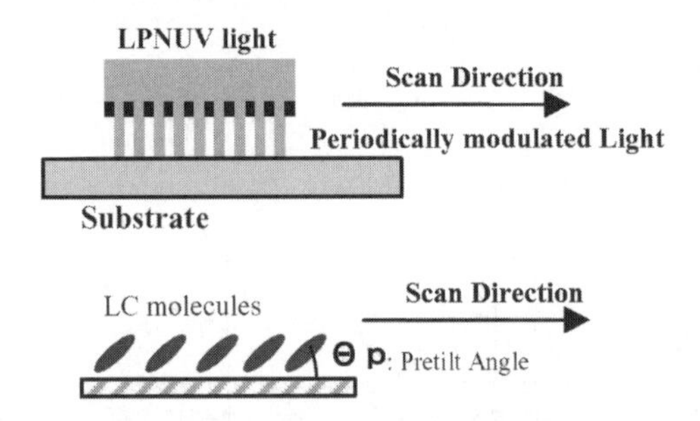

Figure 3.6 Photorubbing method consisting of normal incidence single-exposure method by linearly polarized near-UV (LPNUV) light accompanied with scanning in one direction [13]. Reproduced from M. Kimura, S. Nakata, A. Kumano, and H. Yokoyama, Photo-rubbing method: a single-exposure method to stable liquid-crystal pretilt angle on photoalignment films. *IDW'04 Digest*, p. 35 (2004), IDW Conferences

method [13]. The new system consisted of periodically modulating the intensity of the normally incident, linearly polarized near-ultraviolet light and unidirectionally scanning it on the photosensitive alignment layer (Figure 3.6).

The pretilt appears in such a way that the nematic director is lifted up in the scan direction, similar to the case of conventional cloth rubbing. This 'photoscan' method allows fairly accurate control of the pretilt angle by way of the scan speed and light intensity. The LC pretilt angle was increased to almost 3° at a scan rate of 70 μm/s.

Several attempts have been made to obtain a very low uniform pretilt angle from the LC vertical alignment to enable the vertical aligned nematic (VAN) mode [14–16]. The fabrication of VAN-type LC cells was accomplished by oblique non-polarized light irradiation of thin films of an azobenzene-containing polymer [14]. A low pretilt angle of 3° from the LC vertical alignment was controlled by changing the incident angles of non-polarized UV light and was retained even after heating the cell to 100 °C. The thermal stability of the pretilt angle was attributed to efficient cooperative molecular interactions among mesogenic azobenzene moieties [14]. Another possibility for high-pretilt-angle generation is to use an *in situ* photoalignment method on a homeotropic polyimide layer [15]. Control of the high pretilt angle in NLC with obliquely polarized UV exposure during imidization of the polymer for homeotropic alignment enables a pretilt angle of 85°, which is suitable for VAN LC mode [15].

Photoaligned VAN mode was developed using a commercially available and UV-light-stable-aligning PI material [16]. To photoalign the LC, the polyimide was illuminated by oblique non-polarized UV light. It has been shown that doping the PI with water-soluble sulfonic azo-dye (SD1, Figure 2.4) yields a reliable and perfect electro-optical performance of the VAN mode. This PI/SD1 mixture, used for the preparation of the photoaligning layer, provides the same parameters for the VAN mode response as the conventional rubbing technique. The electro-optical response of the photoaligned VAN mode and the VAN mode obtained by rubbing was very similar despite a certain difference in LC pretilt angles on the substrate. The photoalignment mechanism in the PI/SD1 material includes not only photodegradation, but mostly the pure reorientation of PI/SD1 average molecular absorption oscillators almost parallel to the direction of the obliquely incident UV light (Figure 3.3).

3.2 Generation of Large Pretilt Angles

In many applications, large pretilt angles are needed. For example, large pretilt angles allow the bend mode to be stable at zero bias voltage [17]. It is also needed to achieve the so-called stressed splay–twist (SST) mode, which may be useful for fast LC switching and field sequential color displays [18]. While pretilt angles near 90° for the VAN mode are possible, angles in the range of 10°–80° are difficult to achieve. Nishikawa *et al.* described a method of photoalignment where high pretilt angles were obtained [4, 19, 20]. However, the results were not reproducible. Several other attempts have been made to control the pretilt angle using a single photoalignment layer, with no apparent success. In this section, we shall discuss two other methods of generating large and controllable pretilt angles. The first method makes use of controlled photodegradation of a PI alignment layer. The second method makes use of a nanostructure surface consisting of two types of alignment materials, one of which may be photoalignable.

3.2.1 Generation of Large Pretilt Angles by Controlled Photodegradation

Ordinary PI can be degraded by deep ultraviolet (DUV) radiation. DUV radiation breaks chemical bonds and leads to loss of alignment for the PI. This is a serious problem for displays where the UV intensity from the backlight unit or from the light source is too strong. In particular, for liquid crystal on silicon (LCOS) microdisplays, degradation of the alignment layer is a serious problem, especially for large-size projection displays where light from the metal halide arc lamp light source can be very intense and contains a considerable amount of UV radiation [21, 22].

It turns out that, while for homogeneous alignment PI the degradation leads to loss of alignment, for vertical PI the degradation can lead to a change in the pretilt angle. The change in pretilt angle as a function of the degradation of the vertical alignment PI was first observed by Li *et al.* [23]. Figure 3.7 shows the change in the pretilt angle as a function of DUV light dosage. It can be seen that pretilt angles from 0° to 90° can be obtained by this controlled degradation process. This result is actually quite reproducible. However, since chemical degradation is involved, as in the case of photoalignment using photodegradation, the voltage holding ratio and residual DC (RDC) voltage are not very good. Thus the usefulness of such a process is limited. Nevertheless, it is interesting to note that any pretilt angle can be produced.

This is actually not a surprising result. The physical mechanism of this change in pretilt angle can be explained by noting that the chemical structure of a vertically aligned PI is similar to a homogeneous aligned PI with a side chain that produces the vertical alignment. Apparently the DUV breaks the side chain first, rendering the vertical alignment ineffective. But the main-chain PI still has a homogeneous alignment capability. Thus it becomes a tug-of-war between the vertical alignment and homogeneous alignment portions of the PI.

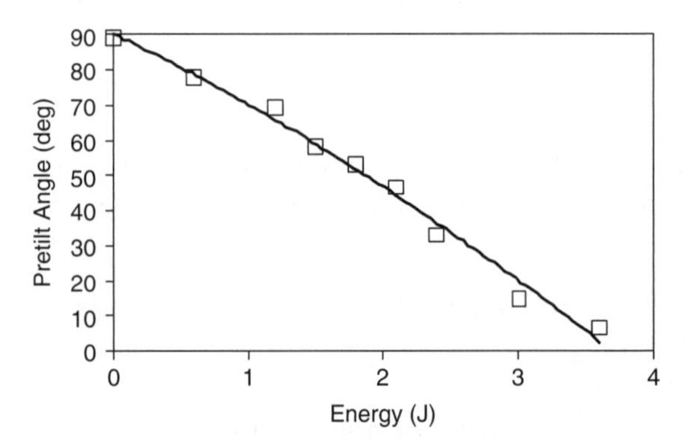

Figure 3.7 Pretilt angle as a function of DUV dosage [23]. Reproduced from a work in progress by Y. W. Li, V. G. Chigrinov, and H.-S. Kwok

A simple heuristic model to understand the alignment property of the PI as a function of degradation can be given. Suppose there is a molecular group within the polymer that gives vertical alignment and another molecular group that gives homogeneous alignment. The surface energy of the LC molecules will be given by

$$W(\theta) = A\sin^2\theta + B\cos^2\theta \tag{3.1}$$

where θ is the tilt angle of the molecules. A and B are coefficients that depend on the alignment strength as well as the concentration of these molecular groups. Equation (3.1) can be rearranged to show the dichotomy of the alignment behavior:

$$W(\theta) = A + (B - A)\cos^2\theta. \tag{3.2}$$

It can be seen that if $A > B$, then the energy will be minimized if the alignment is homogeneous, i.e. $\theta = 0^\circ$. If $A < B$, then the opposite is true, with the homeotropic alignment being preferred. If $A = B$, the alignment is indeterminable. Thus it can be seen that the alignment shows a binary behavior. It is due to the fact that we assume a single homogeneous solution. Thus only one solution is possible, which is either 0° or 90°. If we allow spatial inhomogeneity, as seen in the next section, then it will be seen that intermediate solutions are possible, with pretilt angles between 0° and 90° being produced.

While photodegradation of PI is strictly speaking not a photoalignment process, it does require the application of light. In the next section, we shall describe a true photoalignment process to obtain variable large pretilt angles.

3.2.2 Generation of Large Pretilt Angles by Nanostructured Surfaces

There has always been a need to obtain pretilt angles between 10° and 80°. A new method of generating high pretilt angles using a mixture of vertical (V) and horizontal (H) PI alignment materials has been proposed recently [24]. This method makes use of the nanostructures that are formed on the alignment surface. While on the alignment surface the pretilt angles are either 0° or 90°, away from the surface, elastic energy minimization will necessarily demand that the pretilt angle achieves a uniform angle which is between 0° and 90°. A competition between the two domains therefore results in an intermediate pretilt angle, which is determined by the area ratio of the two types of domains and the elastic constants. The situation is depicted in Figure 3.8. Two domains are shown in that figure. One is nearly homogeneous and the other is nearly homeotropic. It can be seen that, away from the surface, a uniform pretilt of intermediate value is obtained.

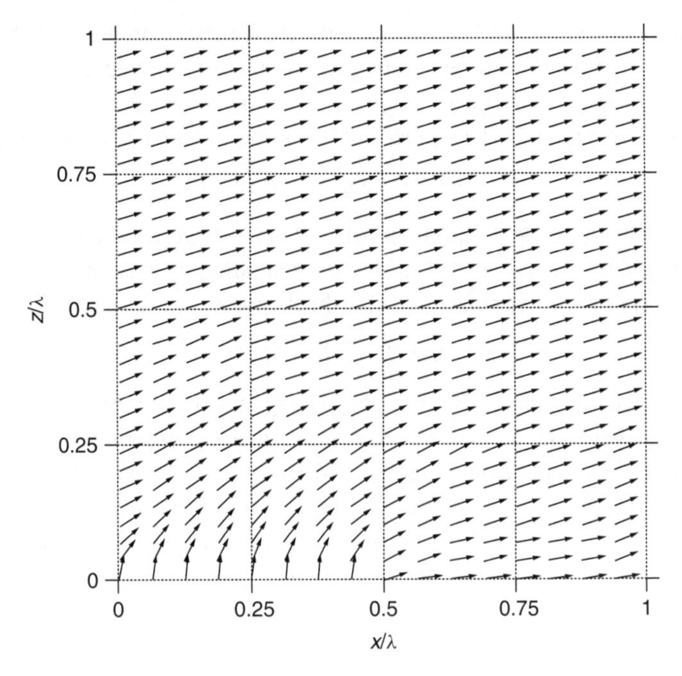

Figure 3.8 Director profile for two adjacent domains with homogeneous and homeotropic alignments [24]. λ is a scale length. Reproduced from F. Yeung, J. Ho, Y. Li, F. Xie, O. Tsui, P. Sheng, and H.-S. Kwok, Variable liquid crystal pretilt angles by nanostructured surfaces. *Applied Physics Letters* **88**, 51910 (2006), American Institute of Physics

In [24], mechanical rubbing of the PI was used to produce alignment. It is therefore interesting to see if photoalignment can be used, which turns out to be the case. Ho *et al.* mixed a homogeneous PI and a homeotropic PI to form an alignment layer as in [24]. DUV light was then used to alter the properties of the PI to vary the resultant pretilt angle. A two-step procedure similar to that in Section 3.1 can be used [25]. The first linearly polarized DUV light exposure is followed by an unpolarized DUV exposure. Figure 3.9 shows the schematic of *in situ* imidization and the two-step DUV exposure setup.

A collimated beam of DUV light from a 1000 W Hg–Xe lamp was polarized by a Glan–Taylor polarizer. During the illumination, the substrate was kept in a vacuum at a constant temperature of 220 °C. This procedure inhibits the polymerization of PI along the polarized DUV light direction, causing a preferential alignment of the LC molecules perpendicular to the polarized light direction. In

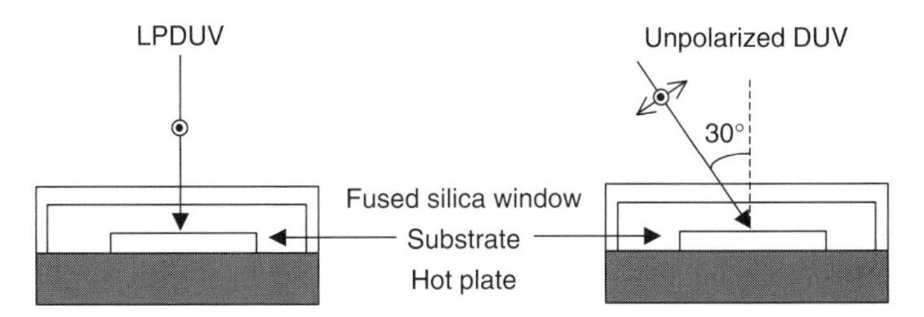

Figure 3.9 Schematic diagram of *in situ* imidization and two-step DUV exposure setup [25]. Reproduced from J. Y. L. Ho, V. G. Chigrinov, and H.-S. Kwok, Variable liquid crystal pretilt angles generated by photoalignment of a mixed polyimide alignment layer. *Applied Physics Letters* **90**, 243506 (2007), American Institute of Physics

order to break the two-fold symmetry of the pretilt angle along the alignment direction, a second unpolarized DUV exposure was required (Figure 3.9). The pretilt angle as a function of concentration of homeotropic PI is shown in Figure 3.10. It can be seen that by varying the homeotropic PI concentration from 30 to 60 wt%, a wide range of pretilt angles can be generated. This result is different from that in [23] in that the variable is the mixing ratio instead of the DUV dosage. It is a more controllable process.

The mechanism of this large-pretilt-angle generation is actually quite interesting. The photoalignment mechanism itself is that of preferential bond breaking. The DUV light breaks bonds in some preferred direction, thus giving the surface an alignment property. However, the control of the pretilt angle is due to an additional mechanism, that of competition between the horizontal and vertical PIs. From the previous section, it can be seen that the pretilt angle of the LC molecules depends on the DUV dosage in the vertical PI layer, due to the process of controlled photodegradation. For pure vertical PI without DUV exposure, an angle of 90° is of course obtained. But if it is subjected to DUV exposure, the anchoring strength will decrease due to degradation of the material. As the vertical alignment side chain is destroyed by the DUV radiation, the pretilt angle will decrease as discussed in Equations (3.1) and (3.2). Now if the alignment surface is uniform and only one pretilt angle is allowed, then it will be either 0° or 90°. But as shown in Figure 3.8, if the alignment surface is spatially inhomogeneous, then the final pretilt angle will depend on the competition between the two types of domains. Thus in the horizontal PI and vertical PI mixture, the final alignment pretilt angle will favor the former if DUV is used. In fact, by controlling the dosage of light, any pretilt angle between 0° and 90° can be achieved as shown in

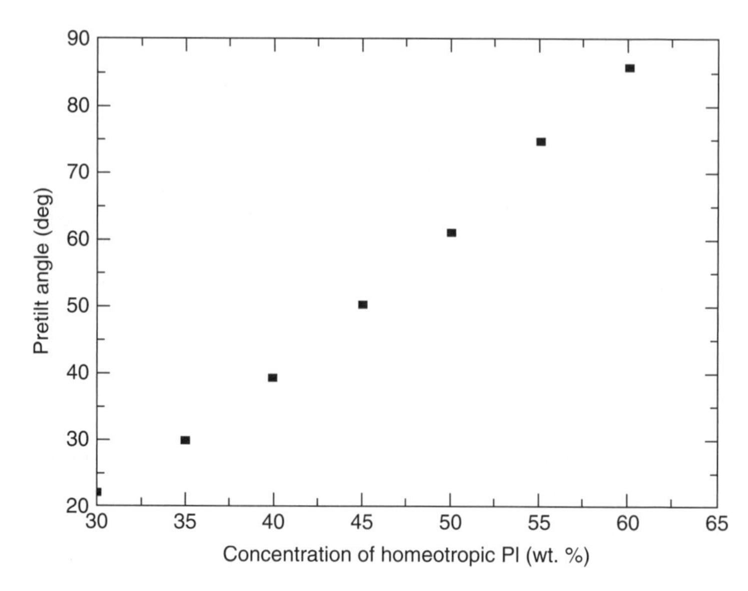

Figure 3.10 Measured pretilt angle as a function of concentration of homeotropic PI [25]. Reproduced from J. Y. L. Ho, V. G. Chigrinov, and H.-S. Kwok, Variable liquid crystal pretilt angles generated by photoalignment of a mixed polyimide alignment layer. *Applied Physics Letters* **90**, 243506 (2007), American Institute of Physics

Figure 3.10. Recently, it was found that the pretilt angle of a rubbed homeotropic PI layer depended on the rubbing strength [26]. High pretilt angles of 40° can be achieved. We suspect that, by strong rubbing, there are regions where most of the side chains have been destroyed, thus making those particular regions have a homogeneous alignment. So the situation may be similar to the case presented in [25] where the vertical PI is destroyed not by mechanical means but by DUV radiation.

This method of generating large pretilt angles using DUV radiation is quite useful and universal. However, since DUV has to be used, the light source is more difficult to control. Further, the use of photodegradation is not desirable as the alignment surface may become contaminated. In a recent experiment, Li *et al.* [23] was able to mix vertical aligned PI with a clean photoalignable polymer to achieve such nanostructures and therefore controlled pretilt angles. This process is still under development and may be a viable method to achieve arbitrary pretilt angles uniformly for a large area.

3.3 Anchoring Energy in Photoaligning Materials

The anchoring energy is one of the most important parameters, which characterizes the LC alignment quality [1]. It usually consists of two parts, azimuthal and polar, that describe how easy it is to change the alignment of the LC director at the surface. The anchoring energy is a part of the LC free energy, needed to realign the LC director from the preferred orientation at the substrate, e.g. from the z axis ($\theta = \varphi = 0$, Figure 3.11) up to the angles θ and φ in the polar and azimuthal direction [1]:

$$W(\theta, \varphi) = 1/2\ W_\theta \sin^2\theta + 1/2\ W_\varphi \sin^2\varphi \approx 1/2\ W_\theta\theta^2 + 1/2\ W_\varphi\varphi^2 \qquad (3.3)$$

where W_θ and W_φ are the corresponding polar and anchoring energies. The value of W_θ and W_φ is a very important characteristic of LC–surface interaction.

To avoid the formation of surface walls and to provide a fast switching 'off' time, the anchoring energy should be sufficiently high, e.g. $W_\varphi > 10^{-4}$ J/m^2, comparable with that of the rubbed polymer surface. This is not really observed for polyvinylcinnamate (PVCN) derivatives [27–29]. The recent measurements also confirm that $W_\varphi \approx 10^{-5}$ to 10^{-6} J/m^2, i.e. about one to two orders of magnitude smaller than the required value [1]. The in-plane sliding mode (IPSL) was proposed in a

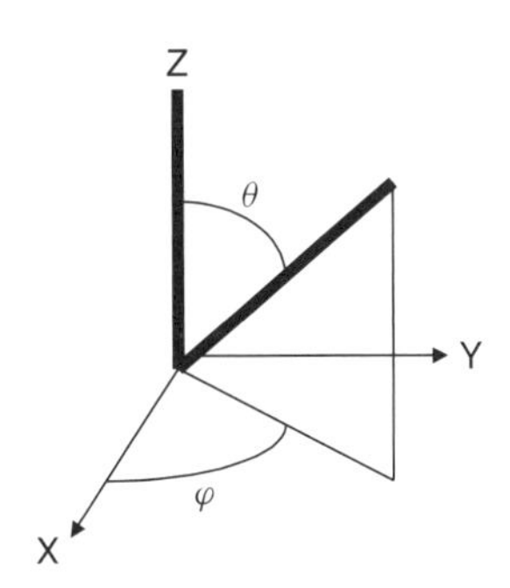

Figure 3.11 The anchoring energy is a part of the LC free energy, needed to realign the LC director from the preferred orientation at the substrate, such as the z axis ($\theta = \varphi = 0$) up to the angles θ and φ in the polar and azimuthal direction: $W(\theta, \varphi) = W_\theta\theta^2 + W_\varphi\varphi^2$, where W_θ and W_φ are the corresponding polar and anchoring energies [1]. Reproduced from V. G. Chigrinov, *Liquid Crystal Devices: Physics and Applications*, Artech House, London (1999)

twisted LC cell with one weakly photoaligned surface using the PVCN derivative, where a significant decrease of the driving voltage was observed, but on account of slowing down the total response time [29]. The azimuthal and polar anchoring energies were measured in a photopolymer containing azobenzene chromophores from 6 to 60% by weight [30]. The azimuthal anchoring strength changed between 10^{-8} and 10^{-6} J/m^2 depending on the concentration of the chromophores and incident exposure energy, while the polar (zenithal) anchoring was typically about 10^{-7} J/m^2 [30]. The polar anchoring strength, measured for the polar monolayers, was an order of magnitude higher [30]: $W_\theta \approx 10^{-4}$ J/m^2. Too small an azimuthal anchoring energy of some photoaligning materials is a reason for so-called sliding (gliding) effects, when the optical axis of the director returns very slowly to its initial state after being disturbed by an electric field in the LC cell [31–33]. Gliding effects are the origin of image sticking and can be avoided only for a sufficiently high LC anchoring energy [34].

The azimuthal anchoring energy can be measured using a substrate with a rubbed PI alignment layer and that with a photoaligned layer assembled with a twisted nematic configuration [35] (Figure 3.12).

The value of the apparent twist angle in the LC cell is lower than the one set by the preferred azimuthal director alignment on the substrates (shown in Figure 3.12

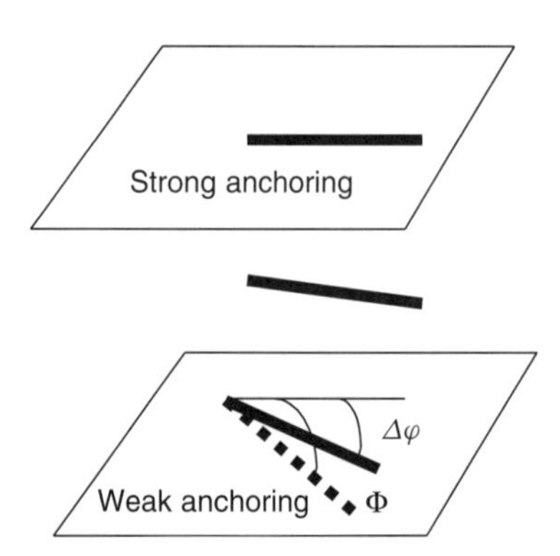

Figure 3.12 The value of the apparent twist angle in the LC cell $\Delta\varphi$ is lower than the angle Φ set by the preferred azimuthal director alignment on the substrates (shown by solid and dashed lines respectively), because the strong azimuthal anchoring substrate affects the director alignment on the weak azimuthal anchoring one [34]. Reproduced from V.P. Vorflusev, H.S. Kitzerow, and V.G. Chigrinov, Azimuthal surface gliding of liquid crystals. Applied Physics Letters **70**, 3359 (1997), American Institute of Physics

by solid and dashed lines respectively), because the strong azimuthal anchoring substrate affects the director alignment on the weak azimuthal anchoring one. The azimuthal anchoring energy of the photoaligning layer W_φ can be calculated from the torque balance equation in the LC cell, provided that the anchoring energy of the rubbed surface is infinite [34]:

$$W_\varphi = 2K_{22}\Delta\varphi/[d\sin 2(\Phi - \Delta\varphi)] \tag{3.4}$$

where Φ, $\Delta\varphi$, K_{22}, and d are the twist angle set by the preferred alignment directions on the two LC cell substrates, the apparent twist angle, the twist elastic constant of LC, and the cell gap, respectively. In [12] the measured apparent twist angle $\Delta\varphi$ was equal to $\Phi = 80^\circ$ with an accuracy of 1°, for the whole range of UV light exposure energies between 1 and 10 J/cm^2, which states that the anchoring energy $W_\varphi > 10^{-4}$ J/m^2, which is comparable with the anchoring energy of the rubbed polyimide layer [1]. As a comparison, in photo-polymerized azo-dye layers, the azimuthal and polar anchoring energies were about 1.5×10^{-5} J/m^2 and 3×10^{-4} J/m^2 for an exposure dose of less than 1 J/cm^2 [12, 36, 37]. Thus the reorientation effect can provide very strong anchoring. This is probably due to the large order parameter that can be achieved with this process. Most of the dye molecules participate in this alignment process, but perhaps only a small fraction in photo-polymerization.

A simple method for the determination of LC polar anchoring energy by electrical measurements was proposed using simultaneous measurements of two homogeneous and homeotropic LC cells to compensate for all the volume effects in the LC bulk and provide a good opportunity to study directly the LC–surface interaction [38]. The method can be applied for LC cells which do not have uniform azimuthal orientation. The polar anchoring energy of the photoaligned azo-dye layer was found to be 3.8×10^{-4} J/m^2, which is comparable with the usual rubbing LC alignment [1].

The thickness of the photoaligned azo-dye layer strongly affects the LC alignment quality. The azo-dye can be used to align the LC even at very small thicknesses. Figure 3.13 shows the measured anchoring energy of the photoalignment layer as a function of the layer thickness [39]. It was found that even a 1 nm thickness (almost a monolayer) could produce uniform LC alignment. However, the anchoring energy was reduced by a factor of two.

The results can be understood by noting that a film with a thickness of 1 nm may not be continuous. In fact it should be in the form of islands. Thus the coverage of the substrate will not be perfect, leading to weaker alignment effects. However,

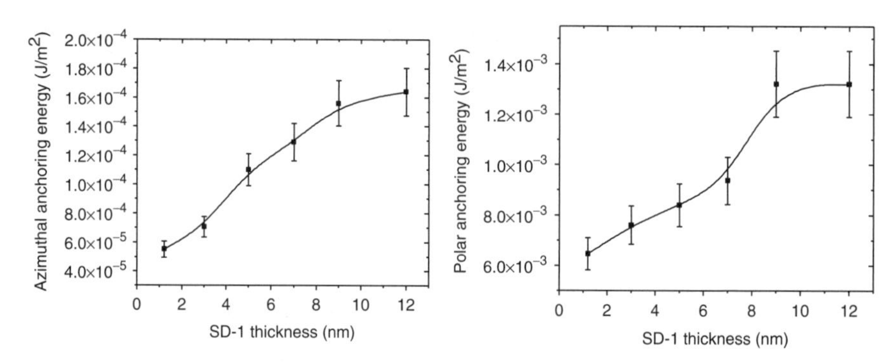

Figure 3.13 LC polar and azimuthal anchoring energy as a function of the thickness of the SD1 photoaligning layer [39]. Reproduced from Vladimir Chigrinov, Hoi Sing Kwok, Hirokazu Takada, and Haruyoshi Takatsu, Photoaligning by azo-dyes: physics and applications. *Liquid Crystals Today* **4**, 1 (2005), IEEE

alignment of the LC does not require a continuous film due to the tendency of the LC molecules to align themselves to minimize elastic deformation energy. The result is that there is still alignment of the LC by the discontinuous layer but the anchoring energy is reduced. Figure 3.13 points to the fact that the alignment is a surface effect. Once a continuous film is obtained, LC alignment can be achieved with good anchoring properties. This is another supporting fact that the azo-dyes are good agents for photoalignment [39].

Another factor, which determines the anchoring energy of the photoaligned layers is the exposure time. In particular, the azimuthal LC anchoring energy of the azo-dye layers can be substantially varied by changing the exposure time (Figure 3.14) [40].

A superthin photoaligned azo-dye layer can also be successfully used for LC photoaligning [41]. This new method includes the formation of very neat 'textile knitwear' by a superthin azo-dye SD1 layer and allows the spin-coating procedure to be avoided. Moreover, the photosensitivity of the azo-dye after photoalignment can be further reduced and the formation of an 'island' azo-dye structure on the rough indium tin oxide (ITO) surface can be prevented due to better adhesion of SD1 molecules. Using this superthin SD1 layer as an alignment agent, sufficiently high polar and azimuthal anchoring energy and perfect LC alignment can be obtained [41]. The method allows perfect LC photoalignment in large or curved cells and is very attractive for mass production. The quality of a photoaligned twisted nematic cell is shown in Figure 3.15 [42].

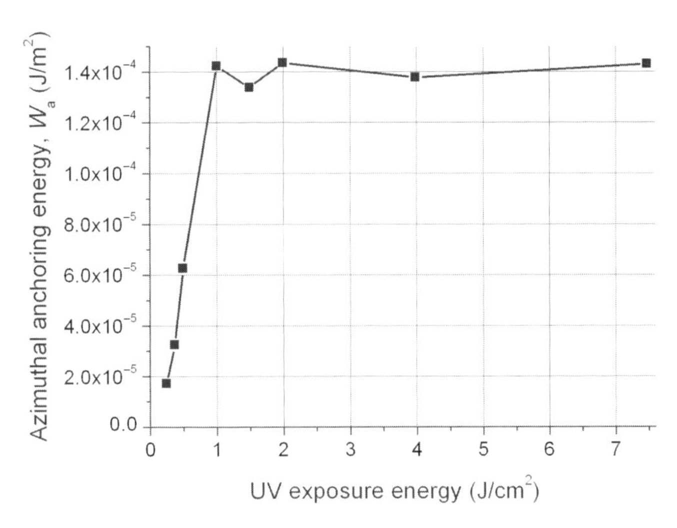

Figure 3.14 The azimuthal LC anchoring energy of the azo-dye layers as a function of the exposure time [40]. Reproduced from Alexei D. Kiselev, Vladimir Chigrinov, and Dan Ding Huang, Photoinduced ordering and anchoring properties of azo-dye films. *Physical Review E* **71**, 061703 (2005), American Physical Society

Figure 3.15 Polarized micrograph of LC alignment on photoaligned polymerized azo-dye film [42]. The size of the picture is 1 mm^2. Both homogeneous and twist LC alignment look perfect between crossed polarizers as the darker and lighter regions. Reproduced from H. Takada, H. Akiyama, H. Takatsu, V. Chigrinov, E. Prudnikova, V. Kozenkov, and H.-S. Kwok, Aligning layers of azo dye derivatives for liquid crystal devices. *SID'03 Digest*, p. 620 (2003), The Society for Information Display

3.4 Stability of Photoaligning Materials: Sensitivity to UV Light

The stability of LC photoaligning materials was a major concern from the very beginning of their development as LC aligning agents [1]. The stable photoaligning materials with respect to UV and IR radiation are photosensitive PIs or crosslinkable materials (see Chapter 2).

The stability of azo-dye photoaligning materials can also be considerably improved by their polymerization. In order to improve their durability against light exposure and moisture, the polymerization of azo-dyes after photoalignment was investigated, synthesizing the azo-dye monomer SDA1 shown in Figure 3.16 [34].

SDA1 was dissolved in dimethylformamide (DMF) and doped with 0.02% of a thermal polymerization initiator V-65 (from Wako Pure Chemicals Industries, Ltd.). The mixture was spin coated onto glass substrates and photoaligned in the same manner as SD1. After the photoalignment, the SDA1 film was heated at 150 °C for 1 h for the purpose of thermal polymerization. LC on the SDA1

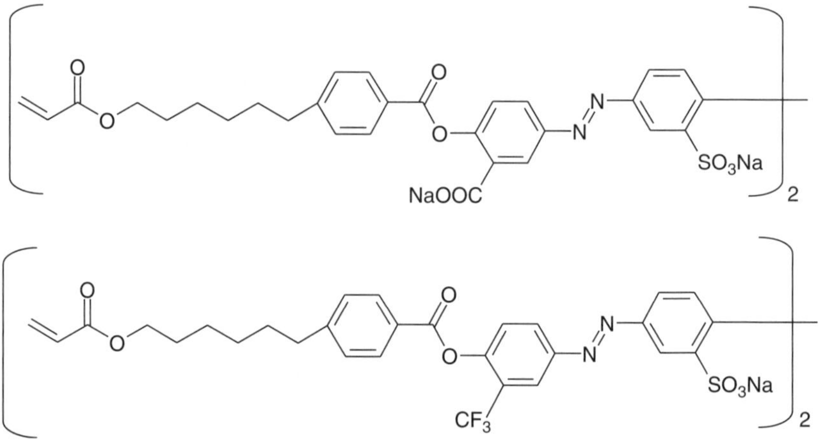

Figure 3.16 Azo-dye monomer SDA1 (upper) and SDA2 (lower) [42]. Reproduced from H. Takada, H. Akiyama, H. Takatsu, V. Chigrinov, E. Prudnikova, V. Kozenkov, and H.-S. Kwok, Aligning layers of azo dye derivatives for liquid crystal devices. *SID'03 Digest*, p. 620 (2003), The Society for Information Display

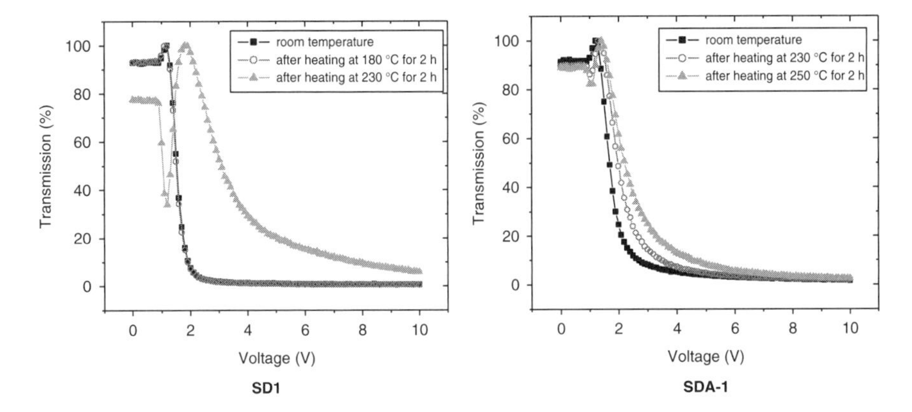

Figure 3.17 Variation of the transmission–voltage characteristics of a photoaligned twisted nematic LC cell after thermal treatment of the cell

film was aligned in the same direction as SD1. The LC alignment properties of the polymerized azo-dye layer were thermally stable up to 250 °C, while SD1 degraded at this temperature (Figure 3.17).

The improvement of durability versus UV light exposure in the LC cell using SDA1 photoaligned film after the thermal polymerization was also confirmed. The dose of UV light at 175 MJ/m^2 can be tolerated by the SDA1 layer, while the SD1 layer is destroyed by this dosage. Moreover, the absorption band of SDA1 is shifted to the UV region, thus making the azo-dye aligning layer more stable to exposure in the visible region (Figure 3.18).

Later the COONa substituent in the formula of SDA1 was replaced by CF_3, thus making the compound more stable against hydrolytic decomposition. The new photo-polymerized azo-dye was called SDA2 (Figure 3.16).

The value of the voltage holding ratio (VHR) was also measured. The measured values for the photoaligned LC cell (>99% at 80 °C) and residual DC voltage (<50 mV) were found to be even better than those for rubbed PI layers. This implies that the azo-dyes studied can be applied as aligning layers in thin film transistor liquid crystal displays (TFT-LCDs), as their purity is high enough and there is no additional ionic impurity from the aligning layers in either the volume or the surface of the LC cell.

The presence of foreign molecules inside the azo-dye film can greatly affect the ease of rotational diffusion during the photoaligning process [43]. In particular,

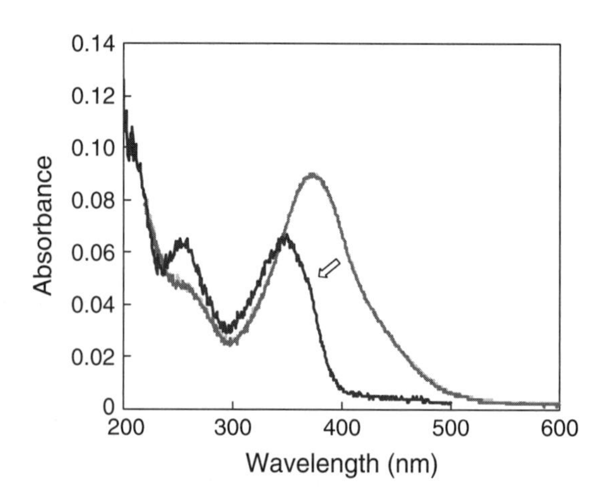

Figure 3.18 The shift of the absorption band of the azo-dye (SD1) to the UV region after thermal polymerization (SDA1)

moisture inside the solid film is detrimental to rotational reorientation due to the polar nature of water molecules. In the standard photoaligning procedure, the alignment layer is first soft baked to drive out most, but not all, of the solvents. SD1 is highly hygroscopic. It can thus absorb moisture, which will affect the photoaligning process by providing more resistance to rotational motion. Recently azo-dye derivatives have been synthesized, which are not soluble in water and thus more resistant to the effect of moisture [43].

Experimentally, it was found that, even with a light dosage of 50 mJ/cm^2, perfect alignment can be achieved in the LC cell by azo-dyes. Certain azo-dye derivatives have been synthesized, which considerably improve the sensitivity of azo-dye SD1, so a very low dosage of UV light is sufficient for perfect uniform alignment of the LC cell: 150 mJ/cm^2 for non-polarized light and 20 mJ/cm^2 for polarized light [43].

We have commented on the stability and sensitivity of the other photoaligning materials in Chapter 2. However, one thing is very important: the sensitivity of the photoaligning material might be considerably improved if it is possible to use the nonlinear regime of the photoalignment [44]. In particular, pulsed irradiation with an appropriate pulse length and energy leads to a stronger increase in birefringence (order parameter) of photosensitive layers, proving that additional thermal and cooperative effects are involved at high excitation densities [44].

3.5 Comparison of the Characteristics of Photoalignment Layers for Different Mechanisms of LC Photoalignment

As follows from Table 3.1, pure reorientation (diffusion) is the most suitable mechanism for LC photoalignment. The commercialization of the materials produced by the DIC Corporation (Japan) is expected soon [45, 46]. The first target should be the photoaligning of monomers and LC polymers for phase retarder applications (Chapter 5). Photoaligning with azo-dye layers will be considered below.

3.6 Various Methods for the Experimental Characterization of Photoalignment Layers

In Chapter 2 we showed that the quality of LC photoalignment and the photoinduced anisotropy in photosensitive layers can be characterized by the measurement of the *in situ* photoinduced birefringence as a function of the exposure time (see Figures 2.7 and 2.8) or by polarized absorption spectra (Figure 2.12). The dichroic ratio $D_{||}/D_{\perp}$ provides the value of the order parameter (Equation (2.15)), which has a maximum value of $S_{max} = -0.5$ in case the molecules are oriented perpendicular to the polarization of the activating light, or $S_{max} = 1$ in case of parallel alignment of the molecules to the light polarization vector. The value of $S = 0$ ($D_{||}/D_{\perp} = 1$) corresponds to a totally random molecular alignment. (The formula (2.15) and all the above arguments are valid only when the absorption oscillator is parallel to the long molecular axis, as in the case of the SD1 azo-dye (Figure 2.4).) If the relative order parameter S/S_{max} is close to 1, this indirectly confirms the high photoinduced anisotropy of the photoalignment film, and accordingly highly uniform LC alignment. We will briefly review here some other methods, which can also be used to characterize the quality of LC photoalignment.

As shown in Figure 2.6, an atomic force microscope (AFM) can provide a high-resolution image to evaluate the uniformity of the distribution and the size of the photoaligned clusters of azo-dye molecules. High-resolution images of individual molecules of the LC material adsorbed on a photopolymer-treated transparent conducting glass plate coated with indium oxide can also be obtained with a scanning tunneling microscope (STM) [47]. The ultrathin layer of the photopolymer poly-vinyl-4-methoxy-cinnamate (PVMC, Figure 2.13) on the conducting glass plate was deposited by the bulk-induced alignment technique [47, 48] (see also

Table 3.1 Characteristics of photoalignment layers for various mechanisms of LC photoalignment (Chapter 2)

Mechanism/ characteristic	Pretilt angle	Anchoring energy	Ionic purity	IR and UV stability	Manufacturability
Cis-trans isomerization	Small [3, 30]	Low [30]		Poor, spontaneous transfer from *cis* to *trans* form	
Crosslinking in photosensitive polymers	Small [2, 5]	Low [2, 28]			Non-uniform exposure is not tolerated
Photodegradation (photoselection)	Any pretilt angle [19, 25]		Degradation of purity due to by-products	High	The exposure time strongly depends on the chemical content of the material
Pure reorientation (diffusion)	Small [12]	High, comparable with rubbing [38]	High, compatible with TFT-LCD [39, 43]	High [37]	Easy – the only problem is moisture for water-soluble azo-dyes [39, 43]

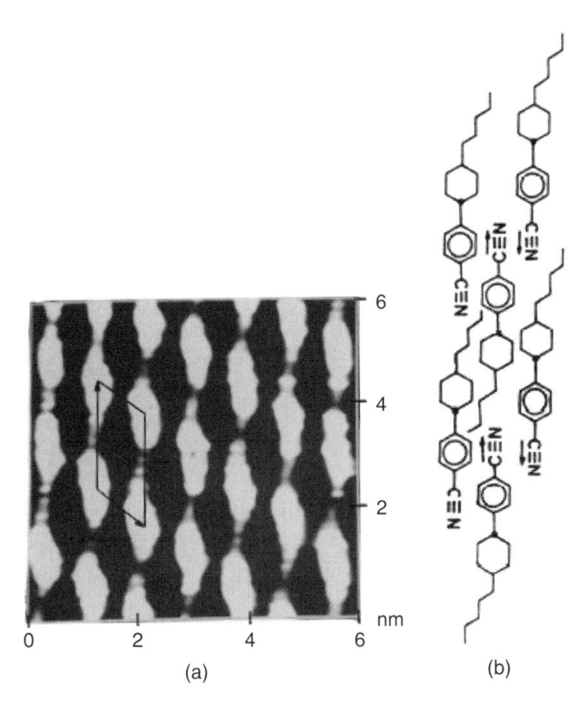

Figure 3.19 (a) A 6 × 6 nm STM image of PCH-5 showing the 2D structure of LC and (b) the packing arrangement of the PCH-5 molecule showing antiparallel ordering (reproduced from S. C. Jain, K. Rajesh, S. B. Samanta, and A. V. Narikar, *Investigation of the interfacial order of nematic liquid crystal on photopolymer coated conducting glass substrates with a scanning tunneling microscope*, Appl. Phys. Lett., **67**, 1527 (1995)

Section 4.9). After evaporation of the solvent and UV exposure, the LC molecules in a layer of PCH-5 (trans-4 pentyl-(4-cyanophenyl)-cyclohexane) from E. Merck exhibit a positional antiparallel ordering, in addition to a high degree of orientational order of the bulk nematic (Figure 3.19).

Polarized electroluminescence (EL) spectra can also be used to characterize the quality of the photoaligned layer. The brightness and polarization ratio of EL is shown to depend on the composition and processing conditions of the alignment layer. Photoalignment provides a non-contact method to achieve macroscopic orientation of the chromophore without mechanical damage. EL with a polarization ratio of 11:1 (Figure 3.20) has been obtained from a fluorene-based nematic network formed by the selective polymerization of diene photoactive

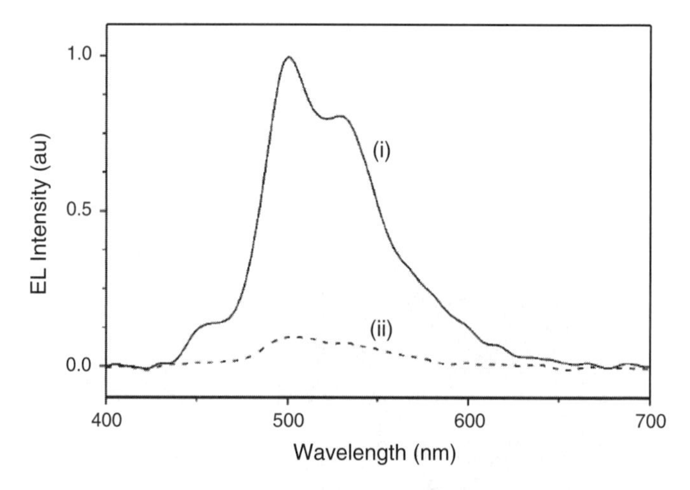

Figure 3.20 Parallel (i) and perpendicular (ii) polarized EL spectrum of a fluorene-based nematic network formed by the selective polymerization of diene photoactive end-groups oriented by a doped coumarin photoalignment layer to promote the transport of holes [49]

end-groups [49]. Surface alignment was achieved using a doped coumarin photoalignment layer (see Figure 2.14), oriented by exposure to polarized UV light. The doping was used to promote the transport of holes conducted to the EL layer. The development of organic light-emitting diodes (OLEDs) now offers the prospect of an alternative lighting source with polarized EL for LCDs with lower power consumption and/or higher brightness. This can be achieved by replacing one of the polarizers and the backlight by a polarized EL light source. More work in this direction is expected.

The surface anisotropy of linearly polarized deep ultraviolet (LPDUV) irradiated PI films was investigated by near-edge X-ray absorption fine structure (NEXAFS) spectroscopy [50]. NEXAFS spectroscopy is known to be a powerful tool for studying orientation in organic molecules and polymers [50]. Core excitations of the second-row elements (C, N, O, etc.) as observed by NEXAFS may show polarization dependence on linearly polarized incident soft X-rays so that the analysis of the angular dependence of each resonance intensity in the spectra gives information of molecular orientation. The NEXAFS studies revealed the preferred molecular orientation at the surface of rubbed polymer films, and concluded that this orientation is the origin of LC alignment on the surface. Matsuie *et al.* [50] examined whether similar outstanding effects occur on the

LPDUV-irradiated PI and established the correlation between the surface anisotropy and the macroscopic alignment capability of PIs. Surface-specific information thus obtained has revealed that the maximum anisotropy is obtained at 0.6 J/cm^2 irradiation, which is in good agreement with the condition of perfect LC alignment capability. For higher doses of radiation, the photoinduced anisotropy of PI film goes to zero, which confirms the concept of the origin of photochemically-induced optical anisotropy, where the photodegradation of PI is an example (see e.g. Figure 2.16).

Certain other methods can also be used to determine the photoanisotropy of photoalignment layers and its correlation with the quality of LC alignment, such as X-ray photoelectron spectroscopy (XPS), that show chemical changes at the surface and interfaces of the substrate materials [51], surface optical second harmonic generation [52], and spectroscopic ellipsometry [53, 54]. In particular, the diffusion model of pure reorientation of azo-dye molecules under the action of an activating polarized light (Section 2.2.1) was recently confirmed by ellipsometric measurements, and no chemical changes were revealed in the UV exposed azo-dye layer [54]. In parallel, spectroscopic ellipsometry allows evaluation of the refractive indices and extinction coefficients of the azo-dye layer in the whole visible range before and after UV exposure (Figure 3.21).

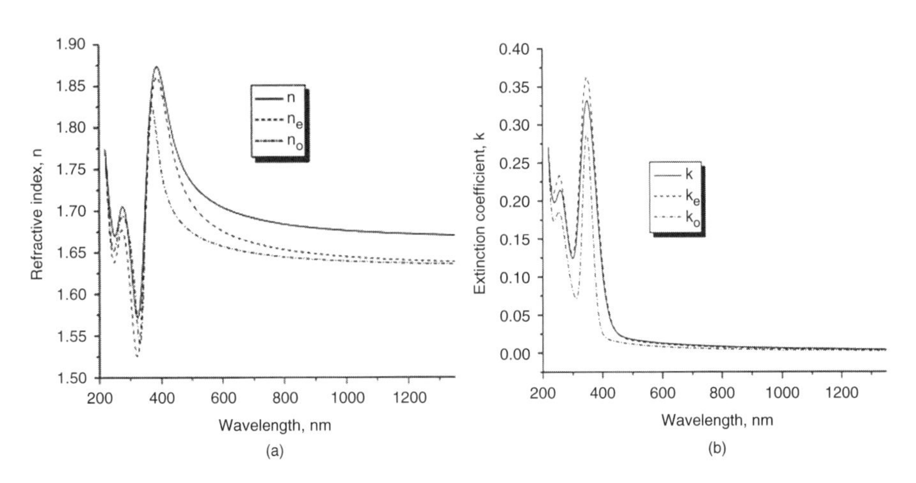

Figure 3.21 Spectral refractive indices (a) and extinction coefficients (b) of photoaligned azo-dye (SD1/SDA2) mixture layers as in the isotropic state (solid line) and after illumination with linearly polarized UV light (dashed and dot–dashed lines) [54]

References

[1] V. G. Chigrinov, *Liquid Crystal Devices: Physics and Applications*, Artech House, London (1999).

[2] A. Dyadyusha, A. Khizhnyak, T. Marusii, V. Reshetnyak, Yu. Reznikov, and W.-S. Park, Peculiarity of an oblique liquid crystal alignment induced by a photosensitive orientant. *Japanese Journal of Applied Physics* **34**, L1000 (1995).

[3] Y. Iimura, T. Saitoh, S. Kobayashi, and T. Hashimoto, Liquid crystal alignment on photopolymer surfaces exposed by linearly polarized UV light. *Journal of Photopolymer Science and Technology* **8**, 258 (1995).

[4] M. Nishikawa and J. West, Mechanism of generation of pretilt angles on polyimides with a single exposure of polarized ultraviolet light. *Japanese Journal of Applied Physics* **38**, 5183 (1999).

[5] M. Schadt, H. Seiberle, and A. Schuster, Optical patterning of multidomain liquid-crystal displays with wide viewing angles. *Nature* **381**, 212 (1996).

[6] D.-S. Seo and C.-H. Lee, Investigation of liquid crystal alignment and pretilt angle generation in the cell with linearly polarized UV light irradiation on polymer surface. *Molecular Crystals and Liquid Crystals* **329**, 255 (1999).

[7] D.-S. Seo and J.-M. Han, Generation of pretilt angle in NLCs and EO characteristics of a photoaligned TN-LCD with oblique non-polarized UV light irradiation on polyimide surface. *Liquid Crystals* **26**, 959 (1999).

[8] D.-S. Seo and J.-H. Choi, Generation of high pretilt angle in a nematic liquid crystal with single oblique polarized UV light irradiation on polyimide surfaces. *Liquid Crystals* **26**, 291 (1999).

[9] D.-S. Seo and H.-K. Kim, Generation of pretilt angle for nematic liquid crystal using an *in situ* photoalignment method on polymer surfaces. *Japanese Journal of Applied Physics* **39**, L993 (2000).

[10] K. Sakamoto, K. Usami, T. Sasaki, Y. Uehara, and S. Ushioda, Pretilt angle of liquid crystals on polyimide films photoaligned by single oblique angle irradiation with unpolarized light. *Japanese Journal of Applied Physics* **45**, 2705 (2006).

[11] B. Park, K.-J. Han, Y. Jung, H.-H. Choi, H.-K. Hwang, S. Lee, S.-H. Jang, and H. Takezoe, Director tilting of liquid crystals on photoisomerizable polyimide alignment layers doped with homeotropic surfactant. *Journal of Applied Physics* **86**, 1854 (1999).

[12] V. Chigrinov, E. Prudnikova, V. Kozenkov, H. Kwok, H. Akiyama, H. Kawara, H. Takada, and H. Takatsu, Synthesis and properties of azo dye aligning layers for liquid crystal cells. *Liquid Crystals* **29**, 1321 (2002).

[13] M. Kimura, S. Nakata, A. Kumano, and H. Yokoyama, Photo-rubbing method: a single-exposure method to stable liquid-crystal pretilt angle on photoalignment films. *IDW'04 Digest*, p. 35 (2004).

[14] S. Furumi, M. Nakagawa, S. Morino, K. Ichimura, and H. Ogasawara, Photogeneration of high pretilt angles of nematic liquid crystals by azobenzene-containing polymer films. *Applied Physics Letters* **74**, 2438 (1999).

[15] Y.-J. Jeon, J.-Y. Hwang, D.-S. Seo, S.-H. Ham, and J.-I. Han, Control of high pretilt angle for nematic liquid crystal on homeotropic alignment layer by in-situ photoalignment method. *Molecular Crystals and Liquid Crystals* **412**, 269 (2004).

[16] V. A. Konovalov, V. G. Chigrinov, H.-S. Kwok, H. Takada, and H. Takatsu, Photoaligned vertical aligned nematic mode in liquid crystals. *Japanese Journal of Applied Physics* **43**, 261 (2004).

[17] Fion Sze-Yan Yeung and Hoi-Sing Kwok, Fast-response no-bias-bend liquid crystal displays using nanostructured surfaces. *Applied Physics Letters* **88**, 063505 (2006).

[18] Yuet-Wing Li, Li Tan, Fion Sze-Yan Yeung, and Hoi-Sing Kwok, Passive-matrix-driven field-sequential-color LCD, *38th Society for Information Display International Symposium, Long Beach, CA*, 154–157 (2007).

[19] M. Nishikawa and J. West, Generation of pretilt angles on polyimides with a single linearly polarized UV exposure. *Proceedings of the SID* **29**, 131 (1998).

[20] M. Nishikawa and J. West, Generation of pretilt angles on polyimides with a single linearly polarized UV exposure. *Molecular Crystals and Liquid Crystals* **329**, 579 (1999).

[21] J. Lu, S. V. Deshpande, E. Gulari, J. Kanicki, and W. L. Warren, Ultraviolet light induced changes in polyimide liquid-crystal alignment films. *Journal of Applied Physics* **80**, 5028 (1996).

[22] W. Oepts, E. Ito, M. Reijme, A. Verschueren, E. Alexander, and C. van der Marel, Degradation of liquid crystal alignment upon high intensity illumination in microdisplays. *International Display Research Conference (Eurodisplay)*, 201 (2002).

[23] Y. W. Li, V. G. Chigrinov and H.-S. Kwok, Unpublished data.

[24] F. Yeung, J. Ho, Y. Li, F. Xie, O. Tsui, P. Sheng, and H.-S. Kwok, Variable liquid crystal pretilt angles by nanostructured surfaces. *Applied Physics Letters* **88**, 51910 (2006).

[25] J. Y. L. Ho, V. G. Chigrinov, and H.-S. Kwok, Variable liquid crystal pretilt angles generated by photoalignment of a mixed polyimide alignment layer. *Applied Physics Letters* **90**, 243506 (2007).

[26] G. P. Sinha, B. Wen, and C. Rosenblatt, Large continuously controllable pretilt from the vertical orientation. *Applied Physics Letters* **79**, 2543 (2001).

[27] V. M. Kozenkov, V. G. Chigrinov, and H.-S. Kwok, Photoanisotropic effects in poly(vinyl-cinnamate) derivatives and their applications. *Molecular Crystals and Liquid Crystals* **409**, 251 (2004).

[28] G. P. Bryan Brown and I. C. Sage, Photoinduced ordering and alignment properties of polyvinylcinnamates. *Liquid Crystals* **20**, 825 (1996).

[29] D. Andrienko, F. Barbet, D. Bormann, Yu. Kurioz, S.-B. Kwon, Yu. Reznikov, and M. Warenghem, Electrically controlled director slippage over a photosensitive aligning mode; in-plane sliding mode. *Liquid Crystals* **27**, 365 (2000).

[30] L. T. Thieghi, R. Barberi, J. J. Bonvent, E. A. Oliveira, J. A. Giacometti, and D. T. Balogh, Manipulation of anchoring strength in an azo-dye side chain polymer by photoisomerization. *Physical Review E* **67**, 041701 (2003).

[31] D. Shenoy, L. Beresnev, D. Holt, and R. Shashidhar, Tuning polar anchoring energy through chemical modification of photodimerized surfaces. *Applied Physics Letters* **80**, 1538 (2002).

[32] S. Faetti, G. Mutinati, and I. Gerus, Measurements of the azimuthal anchoring energy at the interface between a nematic liquid crystal and photosensitive polymers. *Molecular Crystals and Liquid Crystals* **421**, 81 (2004).

[33] S. V. Pasechnik, V. G. Chigrinov, D. V. Shmeliova, V. A. Tsvetkov, V. N. Kremenetsky, L. Zhijian, and A. V. Dubtsov, Slow relaxation processes in nematic liquid crystals at weak surface anchoring. *Liquid Crystals* **33**, 175 (2006).

[34] V. P. Vorflusev, H. S. Kitzerow, and V. G. Chigrinov, Azimuthal surface gliding of liquid crystals. *Applied Physics Letters* **70**, 3359 (1997).

[35] D.-H. Kim, S.-K. Park, S.-B. Kwon, Yu. Kurioz, Yu. Reznikov, O. Tereshchenko, and I. Gerus, Newly developed cellulose-based photopolymer with high anchoring energy and low-image-sticking. *SID'06 Digest*, p. 867 (2006).

[36] V. P. Vorflusev, H. S. Kitzerow, and V. G. Chigrinov, Azimuthal anchoring energy in photoinduced anisotropic films. *Japanese Journal of Applied Physics* **34**, L1137 (1995).

[37] V. G. Chigrinov, H.-S. Kwok, H. Takada, and H. Takatsu, New developments in liquid crystal photoaligning by azo-dyes. *SID'06 Digest*, p. 1253 (2006).

[38] V. Chigrinov, A. Muravski, H.-S. Kwok, H. Takada, H. Akiyama, and H. Takatsu, Anchoring properties of photoaligned azo-dye materials. *Physical Review E* **68**, 061702 (2003).

[39] V. Chigrinov, H.-S. Kwok, H. Takada, and H. Takatsu, Photoaligning by azo-dyes: physics and applications. *Liquid Crystals Today* **4**, 1 (2005).

[40] A. D. Kiselev, V. Chigrinov, and D. D. Huang, Photoinduced ordering and anchoring properties of azo-dye films. *Physical Review E* **71**, 061703 (2005).

[41] X. Li, V. M. Kozenkov, F. S.-Y. Yeung, P. Xu, V. Chigrinov, and H.-S. Kwok, Liquid-crystal photoalignment by superthin azo dye layer. *Japanese Journal of Applied Physics* **45**, 203 (2006).

[42] H. Takada, H. Akiyama, H. Takatsu, V. Chigrinov, E. Prudnikova, V. Kozenkov, and H.-S. Kwok, Aligning layers of azo dye derivatives for liquid crystal devices. *SID'03 Digest*, p. 620 (2003).

[43] H.-S. Kwok, V. G. Chigrinov, H. Takada, and H. Takatsu, New developments in liquid crystal photoaligning by azo-dyes. *IEEE/OSA Journal of Display Technology* **1**, 41 (2005).

[44] V. Cimrova, D. Neher, R. Hilderbrandt, M. Hegelich, A. von der Lieth, G. Marowsky, R. Hagen, S. Kostromine, and T. Bieringer, Comparison of the birefringence in an azobenzene-side-chain copolymer induced by pulsed and continuous-wave radiation. *Applied Physics Letters* **81**, 1228 (2002).

[45] K. Maruyama, Y. Ono, Y. Suzuki, and T. Ikariya, A plastic color TN-LCD using a photo alignment method. *IDW'05 Digest*, p. 141 (2005).

[46] Y. Kuwana, H. Hasebe, O. Yamazaki, K. Takeuchi, H. Takatsu, V. Chigrinov, and H.-S. Kwok, Optimization of photoalignment layer for in-cell retarder. *IDW'07 Digest*, p. 1673 (2007).

[47] S. C. Jain, K. Rajesh, S. B. Samanta, and A. V. Narikar, Investigation of the interfacial order of nematic liquid crystal on photopolymer coated conducting glass substrates with a scanning tunneling microscope. *Applied Physics Letters* **67**, 1527 (1995).

[48] S. C. Jain and H. S. Kitzerow, Bulk-induced alignment of nematic liquid crystals by photopolymerization. *Applied Physics Letters* **64**, 2946 (1994).

[49] A. Contoret, S. Farrar, P. Jackson, S. Khan, L. May, M. O'Neill, J. Nicholls, S. Kelly, and G. Richards, Polarized electroluminescence from an anisotropic nematic network on a non-contact photoalignment layer. *Advanced Materials* **12**, 971 (2000).

[50] N. Matsuie, Y. Ouchi, H. Oji, E. Ito, H. Ishii, K. Seki, M. Hasegawa, and M. Harnikov, UV-photoinduced surface anisotropy of polyimide studied by near-edge X-ray absorption fine structure spectroscopy. *Japanese Journal of Applied Physics* **42**, L 67–(2003).

[51] L. Su, J. West, Yu. Reznikov, K. Artyushkova, and J. Fulghum, XPS characterization of photoalignment using adsorbed dichroic materials. *SID'01 Digest*, p. 1166 (2001).

[52] I. Olenik, M. Kim, A. Rastegar, and Th. Rasing, Alignment of liquid crystals on a photosensitive substrate studied by surface optical second-harmonic generation. *Physical Review E* **61**, R3310 (2000).

[53] C. Ting, H. Akiyama, T. Saitoh, and Y. Iimura, Anisotropy characterization of photoaligned 5CB liquid crystals by generalized ellipsometry. *SID'00 Digest*, p. 159 (2000).

[54] I. Valyukh, H. Arwin, V. Chigrinov, and S. Valyukh, Characterization of the photoalignment material SD1/SDA2 with spectroscopic ellipsometry. *IDW'07 Digest*, p. 391 (2007).

4

Photoalignment of LCs

4.1 Vertical LC Alignment

Vertically aligned LC display (VAN-LCD) has become very popular for LCD TV applications because of the high contrast and wide viewing angle it affords [1]. Vertical photoalignment has been attempted. VAN-LCD can be aligned even with slantwise unpolarized light, which makes this technique very promising for mass production applications [2–6]. At first, the procedure for obtaining VAN mode by slantwise non-polarized UV light using an azobenzene side-chain polymer with subsequent annealing was suggested by Furumi *et al.* [2] The order parameter increases after annealing, but it remains very small. A similar irradiation method using obliquely incident non-polarized light enabled slightly pretilted homeotropic alignment to be obtained using polyimide films. The photodegradation of the commercially available polyimide aligning materials [3, 4] or crosslinking of photopolymers [5, 6] during the exposure of obliquely incident unpolarized light is believed to be the main process which is responsible for the alignment in this case. Unfortunately, the former process results in a decrease of the voltage holding ratio (VHR), which is a very important parameter, especially for TFT-LCD fabrication.

Photoalignment of Liquid Crystalline Materials: Physics and Applications
V. Chigrinov, V. Kozenkov and H.-S. Kwok

Certain success has been achieved in the development of multidomain LCDs, utilizing homeotropic orientation. Two-domain [7] and four-domain [8] photoaligned VAN-LCDs were also fabricated using LC polymer–linear photopolymerized (LCP-LPP) films. The LC director was homeotropic on one substrate, while the other substrate was prepared with reverse bias tilt angles $\theta = \pm 89^{\circ}$ which cause the central director to tilt in opposite directions upon application of the voltage [7, 8]. The concept of LCD master, i.e. a microstructured polarizer which generates polarized light from non-polarized incident light with locally different polarization directions, was proposed to generate a multidomain photoaligned substrate for an LCD cell by one-step exposure [8]. A wide-viewing-angle LCD structure possessing two arrays of photopolymer gratings arranged orthogonal to each other was proposed [9] (Figure 4.1).

In the off-state of the proposed structure, nematic molecules align mostly perpendicular to the cell surface and are reoriented by distorted electric fields at the grating surfaces to make four different domains. The LCD cell shows excellent extinction in the off-state and wide viewing characteristics in the on-state [9].

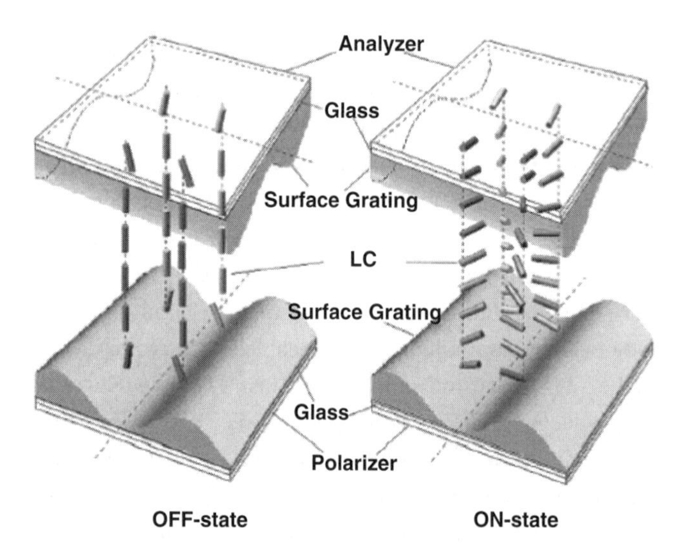

Figure 4.1 The multidomain VAN-LCD based on two arrays of photopolymer gratings arranged orthogonal to each other [9]. Reproduced from J.-H. Park, Y. Choi, T.-Y. Yoon, C.-J. Yu, and S.-D. Lee, A self-aligned multi-domain liquid-crystal display on polymer gratings in a vertically aligned configuration. *Journal of the SID* **11**/2, 283 (2003), The Society for Information Display

However, notwithstanding various efforts, the VAN-LCDs prepared by the photoaligning technique have not yet reached an appropriate quality (response time, contrast ratio) in comparison with conventional LCDs prepared by rubbing technology [2–9].

The application of modern commercial VAN aligning agents with high stability resistivity to UV light in combination with photoaligned azo-dye materials [10] can help to overcome the above-mentioned drawbacks. In an experiment in [11] a commercially available polyimide (PI) for homeotropic alignment was used in combination with an azo-dye for varying the pretilt angle from the homeotropic direction. The composition of 1% of azo-dye in solution with PI was prepared and the photoaligning films were illuminated by oblique incidence non-polarized light. For comparison the aligning film of a pure PI was prepared by the rubbing technique to align the LC molecules in a homeotropic state with some pretilt angle. The measured value of the pretilt angle from the homeotropic alignment was about 1.4°, which is higher than in other photoaligned VAN LC cells (0.53°). Later, pretilt angles of 2°–3° from the homeotropic LC orientation were obtained using a mixture of homeotropic PI and photo-polymerized azo-dye SDA2 (Figure 4.2) [11].

The response times of photoaligned and conventionally rubbed VAN-LCDs were also measured. The results indicate almost no change, with $\tau_{on} + \tau_{off} = 7.8 + 9$ ms for the former and $\tau_{on} + \tau_{off} = 8.1 + 8.9$ ms for the latter. Probably this means that azo-dye as a dopant in PI has increased further the azimuthal anchoring strength in the photoaligned VAN-LCD cells. The high values of VHR for the photoaligned VAN-LCD using the azo-dye/PI composition of 94–96% (close to the conventional

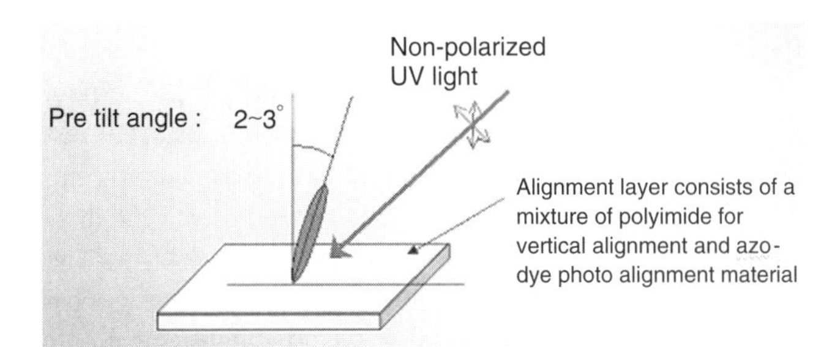

Figure 4.2 Preparation of pretilted vertical alignment for VAN-LCD by a photoalignment method [11]. Reproduced from V. A. Konovalov, V. G. Chigrinov, H.-S. Kwok, H. Takada, and H. Takatsu, Photoaligned vertical aligned nematic mode in liquid crystals. *Japanese Journal of Applied Physics* **43**, 261 (2004), Institute of Pure and Applied Physics

rubbed VAN-LCD with VHR = 98%) testify to this point. The measured value of the contrast ratio between the 'off'- and 'on'-states in all the cases exceeds 1000:1 in monochromatic light ($\lambda = 632.8$ nm). Thus it is concluded that azo-dyes are suitable for the alignment of VAN-LCDs.

Certain success has been obtained in the development of multidomain LCDs, utilizing homeotropic orientation. A four-domain hybrid twisted nematic (HTN) LC cell was fabricated with a small pretilt on one substrate and high on the other, showing a uniform transmission characteristic for wide viewing angles up to 40° [12].

4.2 Twisted LC Photoalignment

The multidomain twisted configuration is suggested to be one of the most important to get wide viewing angles for LCD [1] (Figure 4.3).

The idea is simple: each TN LC cell has the best viewing angles only at a quarter of the whole 2π azimuthal angle, so we have to combine four different TN configurations in one pixel of the LCD (Figure 4.3). The practical realization of multidomain samples, based on a multistep photolithographic process [13], preparation of special polymer layers with two randomly distributed pretilt angles [14], or rubbing only one substrate with a proper doping of the LC with a chiral additive [15], is complicated, so photo-patterning remains one of the most effective techniques to produce a multidomain LCD cell [16]. We have to mention here that multidomain LC cells with almost homeotropic alignment (VAN mode) can be effectively made by the preparation of a special dielectric pattern (protrusions) on one of the electrodes of the LC cell [17].

A dual-domain 65° TN active matrix test display was realized using photoalignment as the non-contact alignment technique [18, 19]. The photoalignment technique was used to create two mutually different LC orientations within one pixel. One substrate in the dual-domain configuration was made by rubbing, while the other substrate was exposed to linearly polarized UV light in two steps to obtain the two subpixels with different orientations in the layer. The configuration was relatively tolerant to variations in the photoalignment process, as it does not need a pretilt on the photoaligned substrates. The shapes of the domain walls between the two twist regions in one pixel on the brightness and contrast of the display were also described [19].

A novel photoaligned TN-LCD cell was fabricated by one-step illumination with oblique non-polarized UV light on an empty cell with azo-dye layers coated

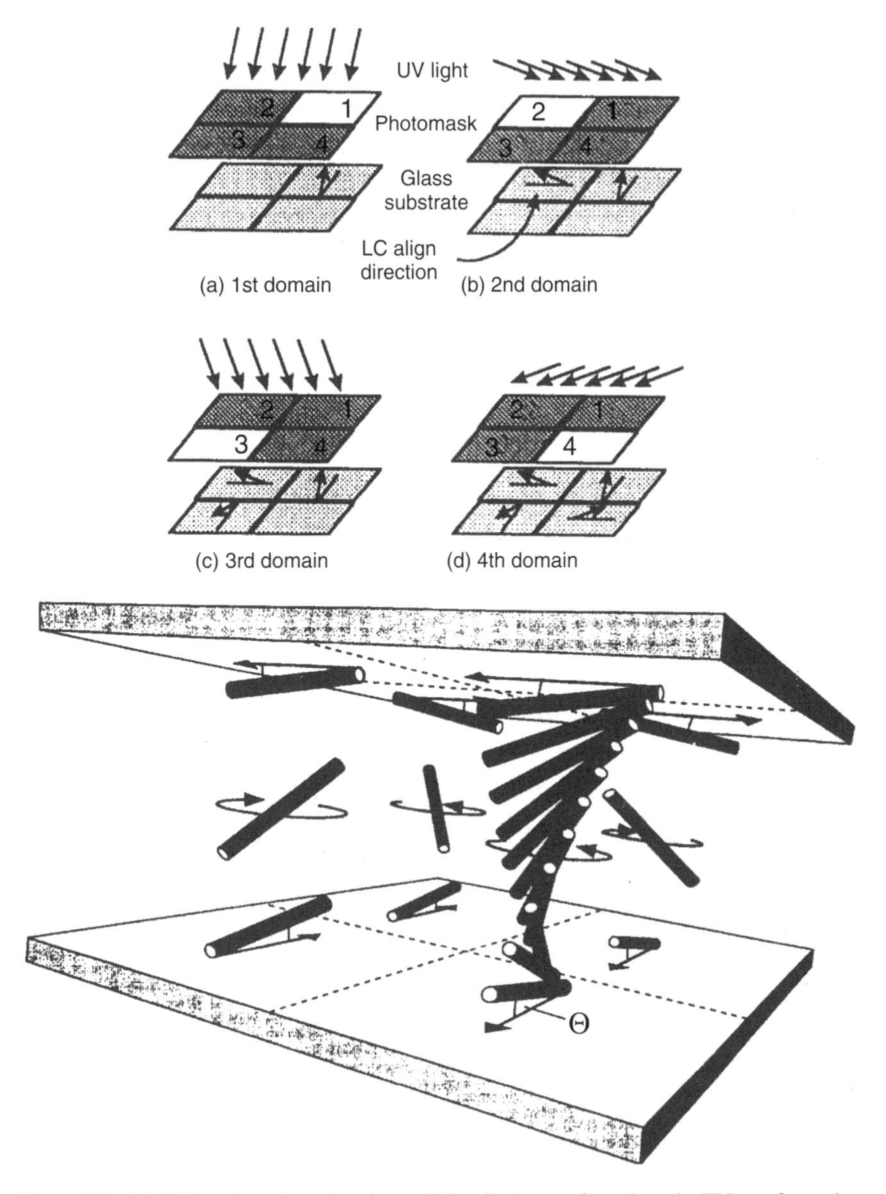

Figure 4.3 Multidomain orientation in an LC cell: lower, four-domain TN configuration; upper, application of a multistep photoaligning technique for its realization [1]. Reproduced from V. G. Chigrinov, *Liquid Crystal Devices: Physics and Applications*, Artech House, London (1999)

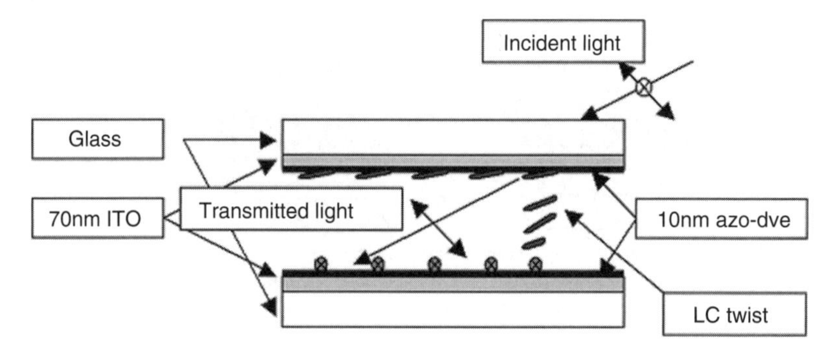

Figure 4.4 UV exposure of empty LC cell by oblique non-polarized UV light. The UV light at a high incidence angle is more p-polarized for the lower azo-dye layer than for the upper layer [20]. Reproduced from D. D. Huang, V. Kozenkov, V. Chigrinov, H.-S. Kwok, H. Takada, and H. Takatsu, Novel photoaligned nematic liquid crystal cell. *Japanese Journal of Applied Physics* **44**, 5117 (2005), Institute of Pure and Applied Physics

on indium tin oxide (ITO) substrates [20]. If the incident angle was sufficiently large ($>75^{\circ}$), the p-polarization of the light was more pronounced for the bottom azo-dye layer in comparison with that for the upper one (Figure 4.4).

Thus, the photoalignment of the azo-dye layer on the bottom substrate was perpendicular to that of the upper one, and LC directors on the top and bottom substrates also became perpendicular. The method provides a simple new way of manufacturing photoaligned TN-LCDs. In-plane switching (IPS-LCD) may benefit from the application of photoalignment technology for twisted LC configurations [21].

4.3 Photoalignment of Ferroelectric LC

Since the beginning of the development of ferroelectric LCDs, the techniques to align them have been a major issue [1]. Anchoring strength, tilt angle, homogeneity, and the required small cell gaps are crucial problems to be solved to render ferroelectric displays capable of manufacture. Ferroelectric liquid crystal (FLC) display cells are extremely sensitive to dust particles that induce nucleation centers for dislocations, especially at a very small cell gap (less than 2 μm) which provides the best optical transmission of the cells [22, 23]. Moreover, buffing nonuniformity very often results in the appearance of large domains with opposite FLC directors (typical dimensions are about 0.5–3 cm). The domain formation leads to poor reproducibility in manufacturing of FLC display devices. The phenomenon of bistability degradation in FLC cells aligned with the ordinary rubbing technique is well known [1]. This is why photoalignment FLC technology, which enables mechanical brushing to be avoided, looks very promising.

A remarkable property of azo-dye aligning layers gives a good chance of providing a high photoalignment quality of FLC using the azo-dye layer. The photoaligned FLC cell appears to be better than that prepared by buffing, if the UV irradiation time of the azo-dye layer is high enough [1]. Perfect bistable switching has been demonstrated in PVA (polyvinyl alcohol) azo-dye photoaligned FLC cells [22]. Bistability degradation was observed even in a photoaligned FLC cell, but it can be easily restored, unlike the case of the FLC cell oriented by the usual rubbing technique. Hybrid photoaligned LC photo-polymerizable polymer (LPP) layers as thin as 12 nm exhibited excellent alignment properties for ferroelectric LCDs [23]. LPP photoalignment technology yields extremely thin, stable alignment layers which result in high contrast ratios without using mechanical treatment. With the alignment problem solved, the inherent advantages of the deformed helix ferroelectric (DHF) effect, namely true grayscale, characteristic response times less than 100 μs, large viewing angle, and low-voltage driving (<4 V for reflective configurations), render DHF-LCDs feasible for a variety of applications [23].

Zigzag-defect-free, surface-stabilized ferroelectric liquid crystal (SSFLC) cells have been prepared using a photodegradable PI irradiated by polarized ultraviolet light (PUVL) at normal incidence to the surface [24]. After irradiation, the FLC molecules were well aligned homogeneously, and the FLC cells showed a uniform texture without zigzag defects, which also indicates a well-aligned layer structure. Zigzag-defect-free alignment may result from the flatter surface, the much smaller and more constant pretilt angles, and the bigger cone angle than those achieved by rubbing. A higher contrast ratio was obtained for a zigzag-defect-free sample obtained by the photoinduced alignment method than by the rubbing method. Faster response times of the order of 300 μs, which are fast enough to realize high-quality moving images with video frame rate, were also obtained [25].

Kobayashi *et al.* have obtained very interesting results in defect-free FLC using photoaligned UV-irradiated PI films [26–30]. It was shown that the defect structure can be suppressed by a proper choice of the FLC pretilt angle and anchoring energy [27]. In particular, a relatively small FLC pretilt angle, large cone angle, and high azimuthal anchoring energy are necessary to promote a uniform C2 state in SSFLC samples [28]. A half V-shaped FLC state, which is useful for active matrix addressing FLC, was promoted in photoaligned FLC cells by using weak azimuthal anchoring energy and polymer stabilization [29]. By doping newly synthesized photocurable monomers in FLC media and performing appropriate photocuring, novel polymer-stabilized FLCDs exhibiting V-shaped switching (PSV-FLCDs) with a response time of 100 μs and a field sequential full-color LCD capable of displaying moving images without blurring and breakup were made [30].

The FLC row addressing response time of $\tau \approx 100$ μs at a voltage pulse amplitude of $U = \pm 15$ V was demonstrated using azo-dye SD1 aligning layers (Figure 2.4) [31, 32]. The thickness of the azo-dye layer should be optimized to get the highest uniformity of the FLC cell (Figure 4.5).

The quality of FLC alignment depended strongly on the azo-dye layer thickness and rather poor aligning texture with irregular focal conic domains was observed with the very thin azo-dye SD1 layer shown in Figure 4.5 [31]. The domains occurred due to the broken SD1 layer, whose evaluated apparent thickness was about 1.2 nm, i.e. approximately two SD1 monolayers were attached to the ITO surface. The texture when the thickness of the SD1 layer was about 3 nm was of a completely continuous film. This uniform texture was a consequence of the perfect uniformity of the azo-dye aligning layer. Increasing the azo-dye layer thickness resulted in the appearance of regular striped domains possessing a periodicity along a normal to the smectic layers of the FLC, as shown in Figure 4.5. These domains were identified as ferroelectric domains in FLC cells [32], which created the modulation of the FLC apparent birefringence. The FLC layers did not become perfectly uniform because of the domains' appearance. The large number of stable states

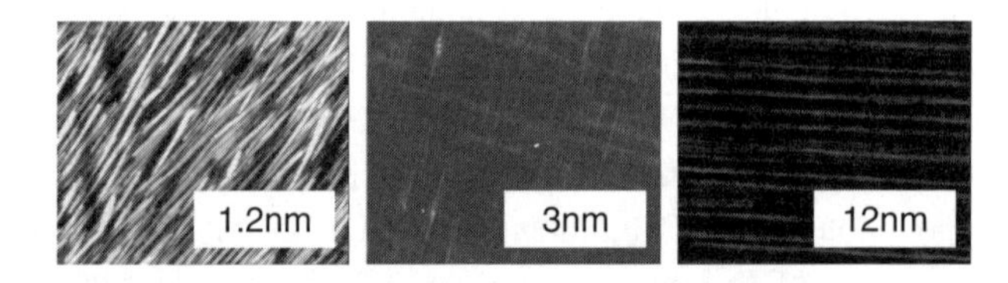

Figure 4.5 Textures of 1.5 mm FLC cells with photoaligned layer azo-dye SD1 of different thickness: 1.2 nm focal conic domains caused by film discontinuity; 3 nm completely continuous film with high uniformity; 12 nm obvious regular striped domains [31]. Reproduced from D. D. Huang, E. P. Pozhidaev, V. G. Chigrinov, H. L. Cheung, Y. L. Ho, H.-S. Kwok, Photoaligned ferroelectric liquid crystal displays based on azo-dye layers. *Displays* **25**, 21 (2004), Springer

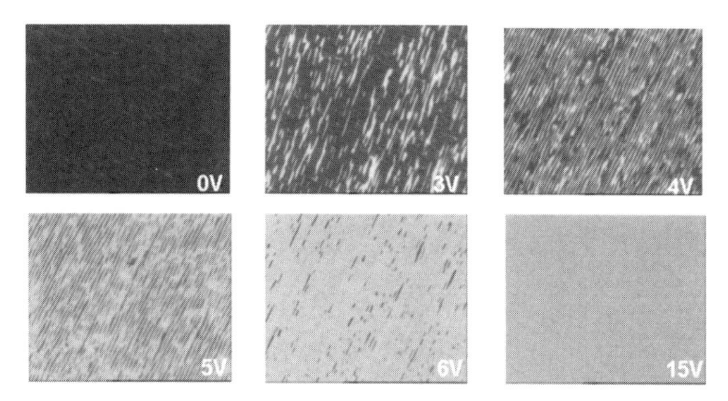

Figure 4.6 Optical appearance of FLC cells with ferroelectric domains dependent on the applied voltage [31–33]. The image area is 350 × 400 μm^2. Reproduced from D. D. Huang, E. P. Pozhidaev, V. G. Chigrinov, H. L. Cheung, Y. L. Ho, H.-S. Kwok, Photoaligned ferroelectric liquid crystal displays based on azo-dye layers. *Displays* **25**, 21 (2004), Springer

of the FLC structure is related to the presence of ferroelectric domains [32]. At a sufficiently high spontaneous polarization and in the absence of helical twist in an FLC layer, modulation of ferroelectric domains by electric field pulses is observed. This modulation results in the appearance of a large number of stable states of the domain structure [33]. However, it is possible that not only ferroelectric domains but also other spatially inhomogeneous structures are responsible for the FLC multistability with an unlimited number of memorized states (Figure 4.6) [31–33].

Thus by switching either the amplitude or the width of the voltage pulse any average gray level of the transmission of the FLC cell can be obtained (Figure 4.7) [33].

Perfect electro-optical performance of the photoaligned FLC display with a memorized grayscale has been demonstrated [31–33]. A prototype of a passively addressed passive matrix FLC display based on the photoalignment technique was developed. Recently, photoaligned bistable FLCs with dichromatic (two-color) [34] and almost achromatic (black/white) [35] displays were realized.

PI photoaligned films were applied for the first time for the alignment of antiferroelectric liquid crystal (AFLC) molecules, which needed planar alignment [36]. LC molecules were aligned perpendicular to the polarization direction of UV light. The AFLC cell showed some defects, but the alignment of the AFLC cell was enhanced by controlling the cooling rate and post-UV treatment.

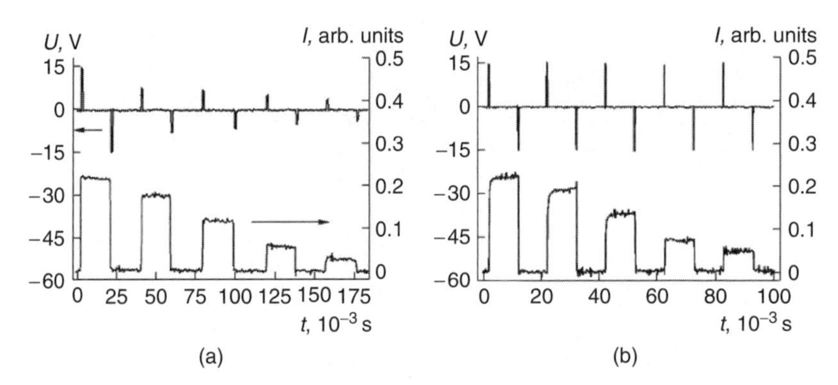

Figure 4.7 Dependences of the change in the light transmission state memorized by the FLC layer on (a) the amplitude of alternating driving pulses 1 ms long and (b) the duration of alternating driving pulses ranging from 250 to 50 μs. The FLC layer thickness is 7 μm; the ferroelectric domain period is 5 μm [33]. Reproduced from E. P. Pozhidaev and V. G. Chigrinov, Bistable and multistable states in ferroelectric liquid crystals. *Crystal Reports* **51**, 1030 (2006), Springer

4.4 Optical Rewritable LC Alignment

The ability of an LCD to change and stabilize its LC structure under light exposure and thus memorize the surface distribution of incident light is very attractive for image recording. Continuous grayscale is one possible benefit of optical 'multistability'. Excellent results were achieved by Yamaguchi *et al.* [37, 38] by controlling the surface distribution of the azimuthal anchoring energy of PVMC film (Figure 2.13). The dose of the exposure energy results in the corresponding azimuthal anchoring energy (Equation (3.4)) and consequently the LC twist angle, which can be visualized between crossed polarizers as the variation of the transmitted light intensity [37, 38]. The approach requires amplitude mask control of the exposure dose and in case of overexposure the image is erased and can hardly be rewritten again. The other development of Yamaguchi *et al.* includes several images written on the LCD using non-polarized UV light, and requires preliminary rubbing of photopolymer crosslinking aligning film materials [39, 40]. The image recording photoaligning technology was successfully applied to an optical security device [41].

The feasibility of a photoswitchable bistable reflective LCD based on an azo-dye doped LC film has been successfully demonstrated [42]. Its bright/dark states can be switched optically, so the device can be written, erased, and rewritten. The image writing was performed with linearly polarized light, but image erasing required circularly polarized light. A dye-adsorbed polymer film with a writing beam intensity of 150 mW/cm^2 yielded a writing time of 60 s. The LC aligning mechanism by surface adsorption of guest azo-dye (dissolvable in LC) was probably too hard to control to enable reproducible rewriting cycles.

The alignment induced on a nematic LC by a photoaligned polymer film with azo-dye side groups has been investigated [43]. The orientation of the LC molecules was manipulated in a reversible manner by irradiating the film with polarized light. The reversibility of the photoinduced alignment may allow recording and erasing of information in an LCD.

All the works mentioned above [37–43] used crosslinking and photodegradation mechanisms (see Chapter 2), so the writing and erasing capabilities of the photoaligning film were very limited (see e.g. Figure 2.15). The true reversible writing–erasing process can be obtained only by diffusion photoalignment in solid films observed in azo-dye layers, such as SD1 (Figure 2.4) [10, 44]. Optical LC alignment can be called *rewritable* when, even in the case of complete decay of the image due to exposure under direct sunlight, the image can be simply restored or changed in an exposure device through a rewriting cycle. Photostability requirements of such an optical rewritable LC cell are significantly diminished as the display unit does not undergo irreversible changes. Optical rewritable (ORW) technology [44] is a modified method of azo-dye photoalignment [10] that possesses traditionally high azimuthal anchoring energy, up to 2×10^{-4} J/m^2, and has a unique feature of reversible in-plane aligning direction reorientation, i.e. rotation perpendicular to the polarization of the incident light. An ORW LC cell consists of two substrates with different aligning materials (Figure 4.8).

One aligning material is optically passive and keeps its aligning direction on one substrate. The other aligning material is optically active and can change its aligning direction being exposed to polarized light through the substrate. In comparison with electrically controlled plastic display, ORW can be significantly thinner and requires *no ITO photolithography* and etching on plastic substrate because *no electrodes* are needed.

Switching and continuous grayscale are achieved by controlling the aligning direction of the photoaligning azo-dye layer, which is insoluble in LC. By this means one can obtain a specified twist angle in the ORW LC cell that corresponds to the transmission level defined by the initial polarizer configuration (Figure 4.8). ORW is very tolerant to cell gap variation and even a 50% change in the cell gap

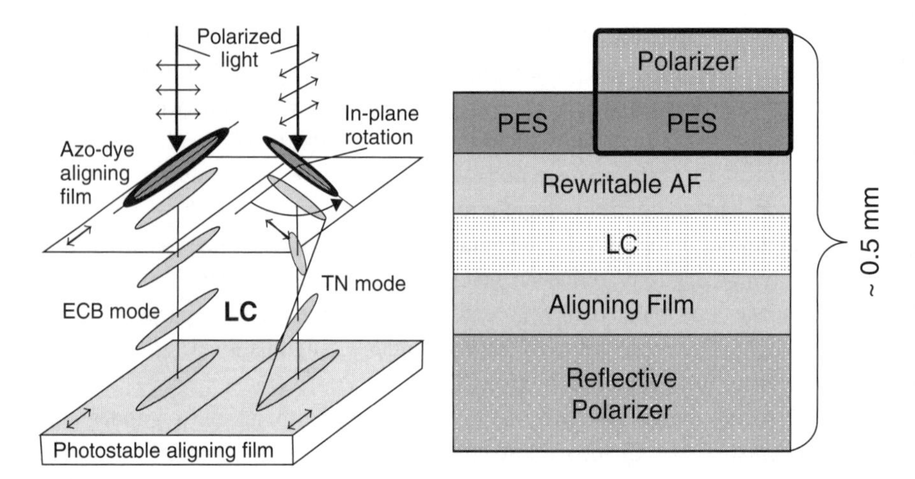

Figure 4.8 Left: operating principle of ORW LC cell. The azo-dye aligning film rotates its aligning direction in-plane keeping perpendicular to the polarization of writing light [44]. The LC follows the top aligning direction switching between homogeneous and twisted states. Right: structure of ORW plastic displays (PES, plastic substrate; AF, aligning film). Reproduced from A. Muravsky, A. Murauski, X. Li, V. Chigrinov, and H.-S. Kwok, Optical rewritable liquid-crystal-alignment technology. *Journal of the SID* **15**/4, 267–273, The Society for Information Display

will not cause a noticeable change in LC transmission value, while achromatic switching of all ORW gray levels can be obtained [44]. In the case of a switch in the twist angle of 0°–70°, the maximum contrast is better than 64:1 and 8:1 for the reflective and transmissive modes. Every transmission level is stable and visualizes information with zero power consumption for a long time.

High reproducibility is achieved due to the saturation of twist angle dependence and careful spectral analysis of the azo-dye. Figure 4.9 presents typical ORW LCD saturation of twist angle dependence on exposure time with 125 mW of polarized light from a high-pressure Hg lamp filtered at a peak of 440 nm.

The polarization angle was selected to obtain saturation for LC twist angles of 12° and 62°. The aligning direction of SD1 rotates continuously in-plane with the saturation in the direction perpendicular to the polarization of the incident light, while the twist angle of the LC cell changes from 12° to saturation at 62° and vice versa.

The average rewriting time dependence on the exposure light intensity is shown in Figure 4.10. At zero intensity level the time should tend to infinity and at high

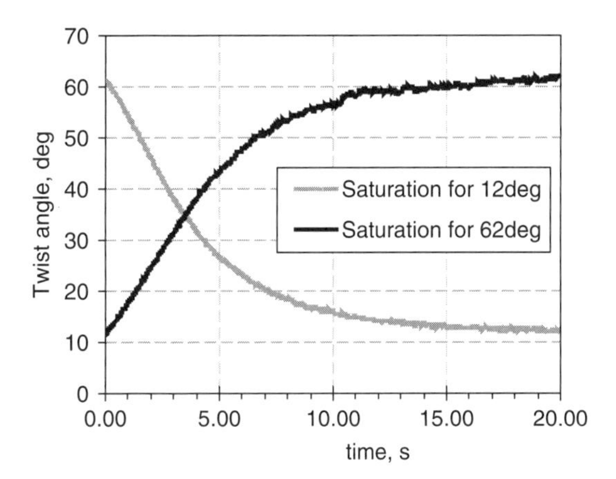

Figure 4.9 In-cell LC twist angle change during exposure process with 125 mW/cm^2 polarized light of high-pressure Hg lamp filtered at a peak of 440 nm [44]. The polarization angle was selected to obtain saturation for 12° and 62° LC twist angles. Reproduced from A. Muravsky, A. Murauski, X. Li, V. Chigrinov, and H.-S. Kwok, Optical rewritable liquid-crystal-alignment technology. *Journal of the SID* **15**/4, 267–273, The Society for Information Display

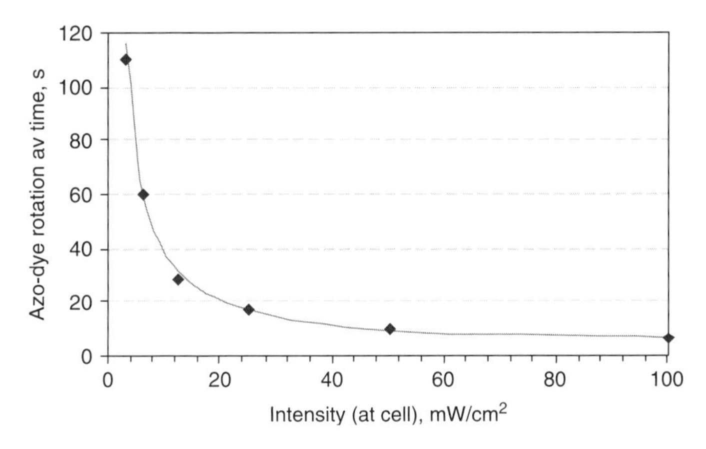

Figure 4.10 Azo-dye rotation (average time) and LC twist angle formation dependence on light intensity [44]. Reproduced from A. Muravsky, A. Murauski, X. Li, V. Chigrinov, and H.-S. Kwok, Optical rewritable liquid-crystal-alignment technology. *Journal of the SID* **15**/4, 267–273, The Society for Information Display

Figure 4.11 Example of rewritable optical memory: written image (left) and negative photomask used to record this image (right) [10]. Reproduced from V. Chigrinov, H.-S. Kwok, H. Takada, and H. Takatsu, Photoaligning by azo-dyes: physics and applications. *Liquid Crystals Today* **4**, 1 (2005), International Liquid Crystal Society

intensity the time decreases. When the exposure intensity exceeds ~20 mW/cm^2 the rewriting time is better than 20 s, which is tolerable for long image-keeping applications. At the same time such an intensity level can be easily reached by a contemporary blue LED.

Optical rewritable LC alignment is a good base for new types of LCDs, especially on plastic substrates, when no ITO films and electrodes are needed (Figure 4.11) [44].

No backlight is required as a reflective-type polarizer is used as the bottom substrate. The image is truly stable and can be written with gray levels, and rewritten a large number of times with high reproducibility of properties.

4.5 Photoalignment with Asymmetric Surface Anchoring

The bistability of bistable π-twisted nematic (π-BTN) LCD is based on asymmetric anchoring of the LC cell, which enables the anchoring breaking effect on one of the surfaces [45]. Using an asymmetric surface anchoring in the LC cell, a truly bistable π-BTN LC cell utilizing photoaligning technology can be fabricated [46]. The strong anchoring surface was achieved by a usual rubbed PI layer. Such a

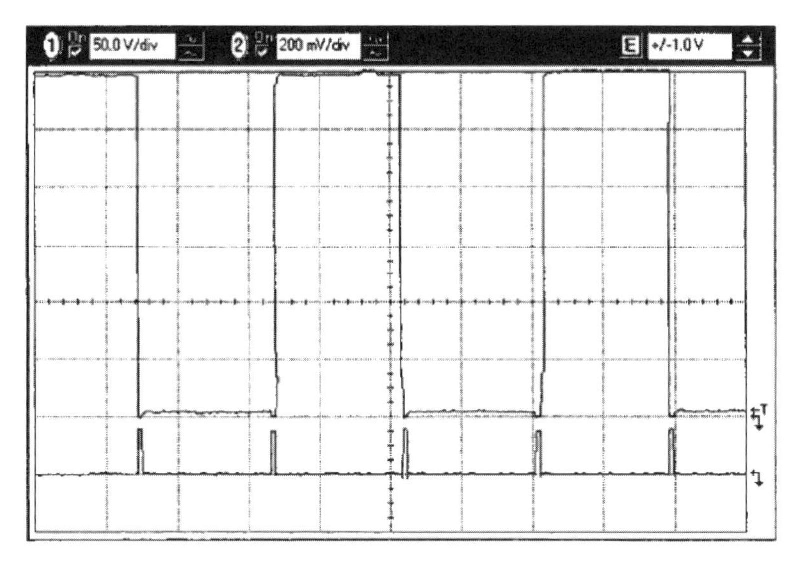

Figure 4.12 Switching behavior of the π-BTN bistable LC cell with asymmetrical anchoring conditions [46]. Top: optical response; bottom: switching pulse train. Reproduced from F. Yeung and H.-S. Kwok, Truly bistable twisted nematic liquid crystal display using photoalignment technology. *Applied Physics Letters* **83**, 4291 (2003), American Institute of Physics

PI layer provides a strong polar anchoring energy of 1.5×10^{-3} J/m^2. The polar anchoring energy of photo-polymerized azo-dye SDA1 (SDA2, Figure 3.14) can be adjusted to $1.5–2.8 \times 10^{-4}$ J/m^2, and azimuthal anchoring to $3.6–5.6 \times 10^{-5}$ J/m^2 by changing the exposure time [45]. This optically bistable LC cell has optimized optical properties and can be switched at a relatively low voltage of 35 V with a pulse width of less than 0.25 ms (Figure 4.12).

The π-BTN LCD was switched between -22.5° and 157.5°, exhibiting a high contrast ratio (>140) and wide viewing angles [46]. A π-BTN LC cell with asymmetrical anchoring conditions was also successfully demonstrated on flexible plastic substrates by using photoalignment technology [47].

4.6 LC Photoalignment on Plastic Substrates

LC alignment on plastic substrates is really an issue, as the usual rubbed PI requires a very high imidization temperature (>250 °C) which is incompatible with the usual plastic substrates, which cannot tolerate such a high temperature [48]. The *in situ* LC photoalignment method on PI surfaces was studied at temperatures from 50 to 120 °C [49]. The electro-optical performance of the *in situ* photoaligned TN LC cell using a polymer substrate was almost the same as that of a TN cell using a glass substrate.

The properties of azo-dye alignment on plastic substrates have also been investigated [50]. Excellent alignment with a high anchoring energy was achieved with an exposure energy of less than 1.0 J/cm^2, which corresponds to the azimuthal anchoring energy of more than 10^{-4} J/m^2. A mixture of azo-dye SD1 (Figure 2.4) with the above-mentioned thermally polymerized azo-dye SDA2 (Figure 3.14) was used for fabrication of the photoaligned LC cell. The LC pretilt angle of about 5° on the plastic substrate was made by a double exposure method (Figure 3.4). The reflectance as a function of applied voltage and response time was measured for both the plastic cell and glass reference cell (Figure 4.13).

A reflective contrast of 8:1 at normal incidence was demonstrated. Perfect uniform LC alignment was achieved on both the plastic and glass substrates [50]. To demonstrate the alignment quality on plastic substrates a nine-digit reflective passive matrix TN-LCD mounted in a smart card was fabricated (Figure 4.14).

To maintain a uniform cell gap of 8 μm (second Mauguin minimum of MLC-6809-000 Merck LC mixture), semi-dry adhesive spacers were chosen. The electro-optical performance of 'the photoaligned plastic display' was very similar to a

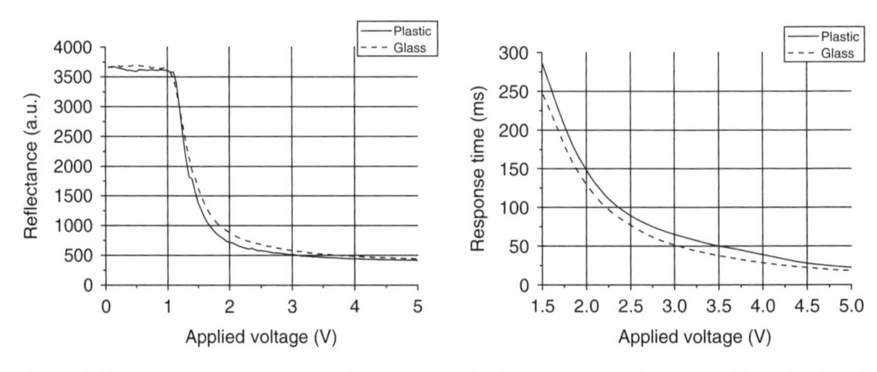

Figure 4.13 Comparison of the LC electro-optical performance in photoaligned TN LC cells on glass and plastic substrates. Reflectance (left) and response time (right) versus applied voltage [50]. Reproduced from J. Osterman, A. Tong, K. Skarp, V. Chigrinov, and H.-S. Kwok, Properties of azo-dye alignment on plastic substrates. *Journal of the SID* **13**, 1003 (2005) The Society for Information Display

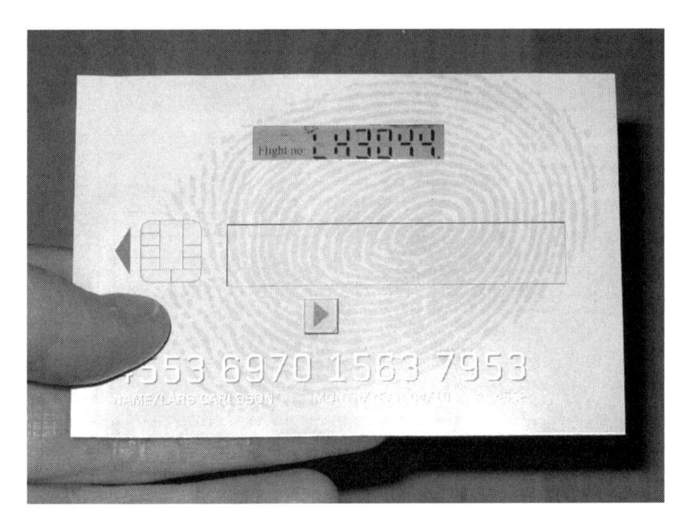

Figure 4.14 Reflective flexible photoaligned TN-LCD mounted in a smart card prototype [50]. Reproduced from J. Osterman, A. Tong, K. Skarp, V. Chigrinov, and H.-S. Kwok, Properties of azo-dye alignment on plastic substrates. *Journal of the SID* **13**, 1003 (2005). The Society for Information Display

common TN-LCD fabricated for comparison by the usual rubbing method on a glass substrate.

A plastic color TN-LCD using an azo-dye photoalignment method has been successfully developed [51]. It was confirmed that LC alignment on a plastic substrate is possible under continuous roll-to-roll processes. The irradiation energies in a double exposure by polarized and oblique non-polarized UV light were 1 J/cm^2 and 0.5 J/cm^2 respectively. The TN-LCD on a plastic substrate was one-sixth thinner and one-tenth lighter than TN-LCD using a glass substrate. A photoaligned plastic LCD prepared by a continuous roll-to-roll coating and orientation process is shown in Figure 4.15.

4.7 Photoalignment on Grating Surface

LC photoalignment on a grating surface is an interesting case of both homogeneous and periodic LC alignment in the plane of the surface [52–59]. A novel LC surface alignment technique, which is a homogeneous LC photoalignment parallel to microgrooves created by optical interference in azopolymer material, has been

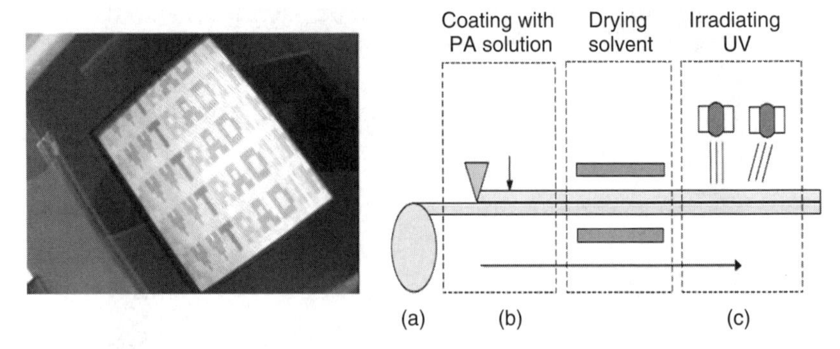

Figure 4.15 Left: plastic color LCD photoaligned by azo-dyes; right: a continuous roll-to-roll process of coating with photoalignment (PA) solution and orientation treatment by azo-dyes [51]

reported [52]. The microgrooves were made directly on the inner surfaces of a fabricated empty cell. The azopolymer film thickness was about 100–200 nm. The writing beam was formed by expanding and collimating a 380 mW/cm^2 circularly polarized argon laser beam at 488 nm. The wavelength of the laser λ as well as the angle of incidence θ of the writing beam to the sample holder can be adjusted to change the spacing ($\lambda/2 \sin \theta$) of the interference pattern. The microgrooves with a sinusoidal profile were inscribed on the polymer film with a spacing of 376 nm and a depth of 80 nm (Figure 4.16).

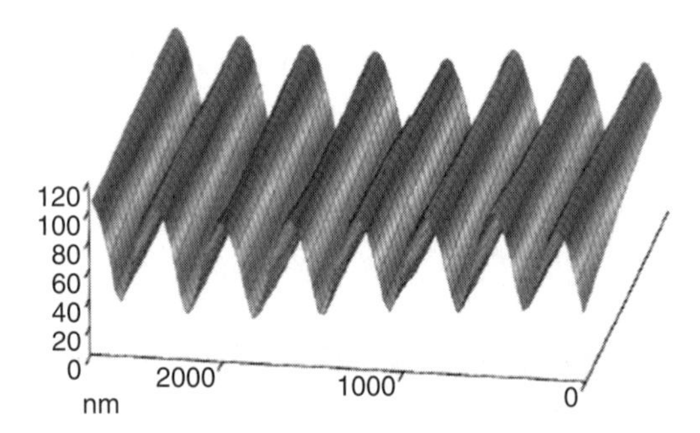

Figure 4.16 Surface relief grating created in a photoalignment surface by the interference method [52]. Reproduced from X. Tong, A. Natansohn, and P. Rochon, Photoinduced liquid crystal alignment based on a surface relief grating in an assembled cell. *Applied Physics Letters* **74**, 3791 (1999), American Institute of Physics

The azimuthal LC surface anchoring energy was more than 10^{-4} J/m^2, which is comparable with rubbed PI films.

A polymer film surface modified by laser illumination has been used as a command surface for nematic LC [53]. The relief was created in a holographic process with a spatial period of 500–1500 nm. Azobenzene-containing polyurethanes were taken as the material of the film. A relief grating inscribed by laser illumination onto the surface of these polymers induced a reorientation of the LC director from its initial homeotropic alignment. The generated surface relief was sinusoidal with an LC anchoring energy for homogeneous alignment of 10^{-6} J/m^2. A surface relief grating was generated in a photopolymer layer deposited on a glass substrate [54]. Such substrates then orient LCs in contact with them resulting in a homogeneous LC alignment. Spatial molecular orientation of the LCs due to steric factors fits the experimental results and is likely to be the dominant aligning mechanism. Homogeneous LC alignment on periodic gratings made by interfering laser beams was also reported in azopolymers containing aromatic azo groups [55] and PI films [56].

A planar periodic alignment in a nematic LC cell has been created by using a command layer of azo-dye molecules directly deposited on the cell substrates and exposed to two interfering laser beams of opposite circular polarizations [57]. A light polarization grating to induce photoisomerization in a pure azo-dye layer deposited onto substrates of an LC cell served as a command layer to create the corresponding LC alignment. As a result a permanent polarization grating was formed in the LC cell (the diffraction from the azo-dye command layer is negligibly small). The empty cell was exposed to spatially modulated polarized light patterns. These polarization gratings were formed using a holographic setup with interfering beams of orthogonal states of polarizations (Figure 4.17).

The continuous wave (CW) Ar^+-ion laser beam (operating at 488 nm) was expanded by a beam expander (BE) to ensure an homogeneous exposure over the whole examined region (the beam diameter was about 6 mm). A beam splitter (BS) was used to obtain two symmetrically incident ‘interfering’ beams. The angle between these beams was adjusted to provide a spatial modulation period $\Lambda = 15$ μm. Glan prisms (PG) and wave plates ($\lambda/2$ and $\lambda/4$) were used to obtain, for example, right circularly polarized (RCP) and left circularly polarized (LCP) beams of equal intensities. The total average power density of the recording beams on the sample surface was 350 mW/cm^2 and the exposure time was 10 min. A plane-polarized CW He–Ne laser beam (operating at 632.8 nm) was used as a probe for real-time and post-exposure monitoring of diffraction gratings formed

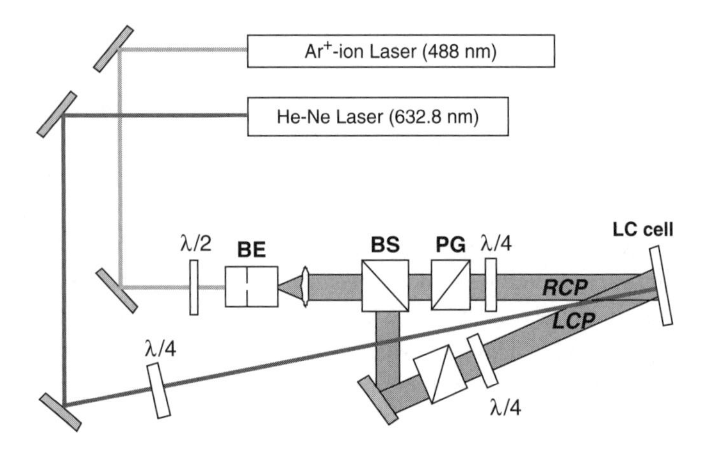

Figure 4.17 Experimental setup for forming the polarization grating in photoaligning layer SD1: λ/2, half-wave plate for power balance between interfering beams; BE, beam expander; BS, beam splitter; PG, Glan prism; λ/4, quarter-wave plates [57]. Reproduced from V. Presnyakov, K. Asatryan, T. Galstian, and V. Chigrinov, Optical polarization grating induced liquid crystal micro-structure using azo-dye command layer. *Optics Express* **14**, 10558 (2006), Optical Society of America

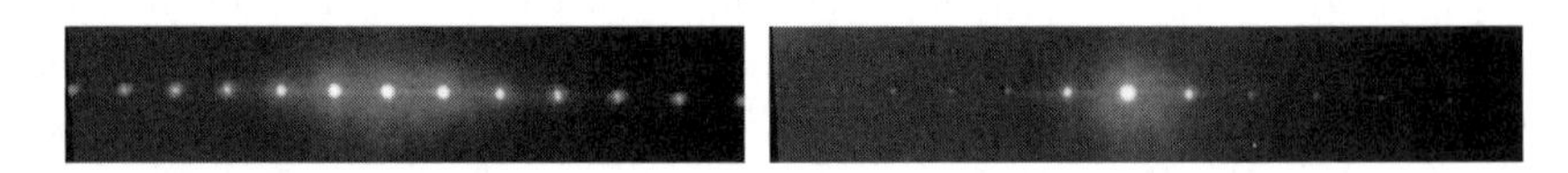

Figure 4.18 Diffraction patterns from LC polarization grating for different polarizations of the probe beam and voltages. Left: $U = 0$ V, linear polarized (p-polarization) probe; right: $U = 15$ V, linear polarized (p-polarization) probe [57]. Reproduced from V. Presnyakov, K. Asatryan, T. Galstian, and V. Chigrinov, Optical polarization grating induced liquid crystal micro-structure using azo-dye command layer. *Optics Express* **14**, 10558 (2006), Optical Society of America

in the LC cell. There was no observable diffraction from empty cells. However, strong diffraction was observed after filling such exposed cells with LC and could be strongly controlled by an applied voltage (see Figure 4.18).

The nature of modulation of the LC director in the cell has been studied using microscope textures of fabricated gratings (see Figure 4.19). Such a structure of LC alignment reproduces the azimuthal distribution of azo-dye molecules in the command layer [57].

Figure 4.19 Micrographs of LC polarization grating [57]: left, between crossed polarizers; right, without analyzer. Reproduced from [57], Optical Society of America

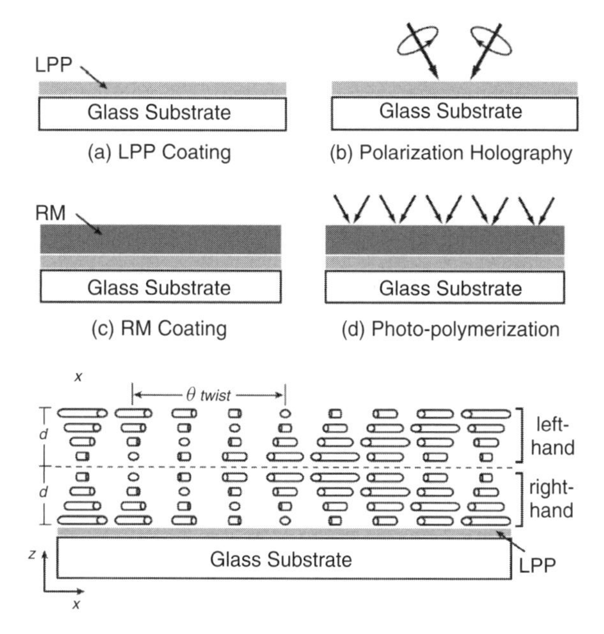

Figure 4.20 Fabrication of an achromatic polarization grating using photoalignment [58]. Reproduced from C. Oh and M. J. Escuti, Achromatic diffraction using reactive mesogen polarization gratings. *SID'07 Digest*, p. 1401 (2007), The Society for Information Display

A periodic boundary condition of LCs using an LPP surface layer [7] has been fabricated using four basic steps (Figure 4.20) [58]: first, a thin layer of LPP is coated on a glass substrate; second, the substrate is exposed to two coherent beams from a laser with orthogonal circular polarizations at a small angle, leading to polarization interference with a constant intensity [59] (see e.g. Figure 4.17); third, the reactive mesogen (RM) mixture is coated on the photoalignment layer and aligned according to the surface pattern; finally, the RM layer is photo-polymerized with a blanket UV exposure to permanently fix the large structured optical anisotropy.

The second RM layer was deposited directly on top of the first, and was subject to the same thickness and twist condition, but with an opposite handedness (Figure 4.20). Finally the achromatic polarization grating (PG) exhibiting only three diffracted orders (0, ± 1) was produced [58]. The achromatic PG offered the high efficiencies of thick (Bragg) gratings over nearly the entire range of visible light. When used as an optical element of displays, the achromatic PG can be implemented as a switchable LC grating for modulator applications, by allowing one of the grating twists to be implemented as a switchable nematic LC and by placing the entire structure between the substrates with electrodes [58].

4.8 Photoalignment of Lyotropic and Discotic LCs

Amphiphilic dye molecules such as sulfonated chromogens can be used in place of PVA iodine for the preparation of thin polarizing films [60]. These molecules are usually water soluble and at a wide range of concentrations, temperatures, and pH values they self-assemble and stack up to form hexagonal complexes. This is usually referred to as the lyotropic LC phase [1]. It can be orientated, by mechanical shear stress, so that the lyotropic liquid crystal (LLC) is orientated parallel to the shear flow direction (Figure 4.21).

However, there are problems associated with shear-flow-induced alignment of LLCs. The first problem is visual defects, which are in the form of horizontal stripes of up to ten micrometers wide. The defects occur at the boundary between areas with different molecular orientations, which are clearly seen in polarized light as vertical bands. These are due to the non-uniform LLC flow gradients, which are sensitive to the shear speed, LLC viscosity, and temperature. The second problem is particulate contamination, which is transferred from the substrate to the rollers and back onto another substrate (Figure 4.21).

LLC photoalignment was first published in [61, 62]. An azo-dye layer AD-1 of about 0.1 μm thickness was spin coated on a glass substrate and illuminated with a shadow mask using polarized UV light to form the photoalignment layer

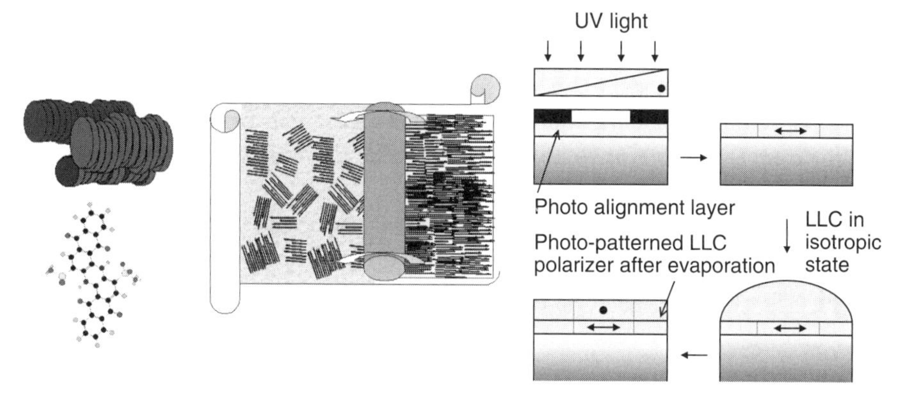

Figure 4.21 LLC photoalignment: left, by mechanical shear stress [60]; right, by photoalignment [61]. Reproduced from V. M. Kozenkov, W. C. Yip, S. T. Tang, V. G. Chigrinov, and H.-S. Kwok, Thin photo-patterned internal polarizers for LCDs. *SID'00 Digest*, p. 1099 (2000), The Society for Information Display

(Figure 4.21). In the second fabrication stage, a few drops of an isotropic LLC solution were dispersed onto the photoalignment layer. When the solvent had evaporated, the LLC was oriented preferentially along the photoinduced axis, where the absorption axis of the LLC was parallel to that of the photoalignment layer. A patterned LLC polarizer demonstrated on the basis of photoaligned LLC will be considered below in Chapter 5 (Figure 5.7).

The surface assisted photoalignment control of LLC was also studied in [63]. Columnar LLC structures of an aqueous solution of a water-soluble dye were aligned on the photosensitive polymer command surface perpendicular to the electric vector of the UV light, since the molecular axis of the dye aligns in the orthogonal direction with respect to the columns consisting of stacked dye molecules (Figure 4.21). The photoaligned LLCs of the dye were stable to heat and light so that stable photoimages were obtained, even though the system contained 90 wt% or more of water, as a result of the transfer of latent images formed by linearly polarized light irradiation of azobenzene polymer films to an LLC layer. The photoalignment control of LLCs is of great significance, since oriented dye molecular films may be fabricated by the evaporation of water from the LLCs and are applicable to micropatterned polarizing elements for stereoscopic and other types of LCDs [1].

As mentioned above, LLCs, which are columnar stacks of discotic molecules, can be photoaligned, thus forming a thin polarizing layer [61–63]. However, the thermotropic discotic LC can also be ordered by a photoalignment method [64–66]. A thin film of 25 nm thickness has been prepared on a fused silica surface from a photosensitive polymer with a side-chain p-cyanoazobenzene photosensitive group [64] (Figure 4.22).

The film was exposed to obliquely incident non-polarized (incidence angle $\theta_i = 45^\circ$) or normally incident polarized light ($\lambda = 436$ nm, $E_{exp} = 3$ J/cm^2) to provide oblique or normal alignment respectively of disk-like molecules (Figure 4.22).

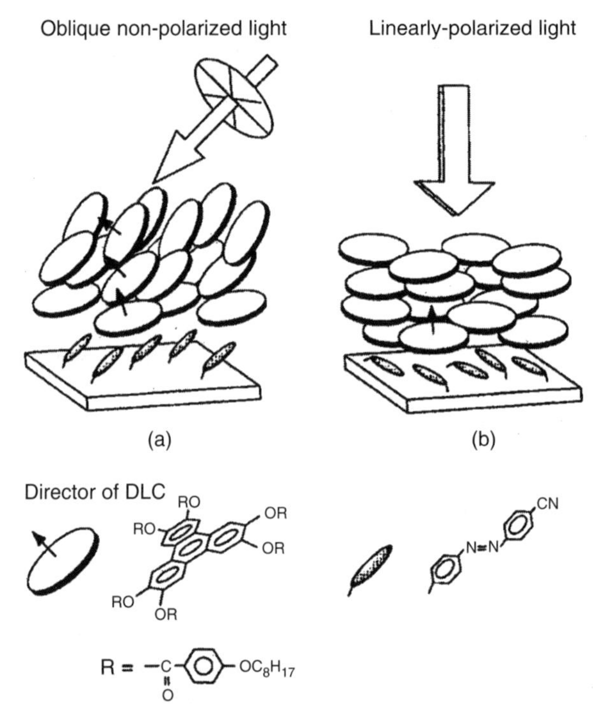

Figure 4.22 Tilted (a) and homeotropic (b) alignment of discotic LC (DLC) molecules (disks) on a thin film of azobenzene units (rods) exposed to oblique irradiation with non-polarized light and normally incident polarized light respectively [64]. Reproduced from K. Ichimura, S. Furumi, S. Morino, M. Kidowaki, M. Nakagawa, M. Ogawa, and Y. Nishiura, Photoinduced orientation of discotic liquid crystals. *Advanced Materials* **12**, 950 (2000), Wiley-VCH

The order parameter and consequently, the alignment quality, increase by the subsequent annealing treatment of the azo-dye film [64]. Similar results of discotic alignment by a PVCN film, crosslinked during the UV exposure, have also been reported [65]. Obliquely irradiated non-polarized light ($\lambda = 313$ nm, $\theta_i = 45^\circ$) was used for the purpose as in a previous case with an exposure energy $E_{exp} = 10$ J/cm^2. Photoalignment of thermotropic DLCs will help to design the new optical retarders for TN-LCDs with wide viewing angles [1]. Surface photoalignment and micropatterning of thermotropic DLCs on the surface of photoirradiated azobenzene polymer films has also been reported recently [66].

4.9 Other Types of LC Photoalignment

Photonic crystal fiber is a glass or polymer fiber with an array of microscopic air holes running along its length. The waveguide properties of such a fiber can be controlled by introducing an additional material into the air holes [10]. LC is suitable for that purpose because its refractive index can be easily tuned by an electric field or by temperature. The technique of photoconfigurable alignment of LC in glass microtubes and in photonic crystal fibers has been developed [67]. Figure 4.23 shows a glass tube with an inner diameter of 4 μm treated with the photoaligning azo-layer SD1 and filled with uniformly oriented nematic LC without point defects or linear disclinations.

The order parameter S of the LC has been obtained from Fourier transform IR spectroscopy data and has demonstrated good alignment quality ($S = 0.63$ [67]).

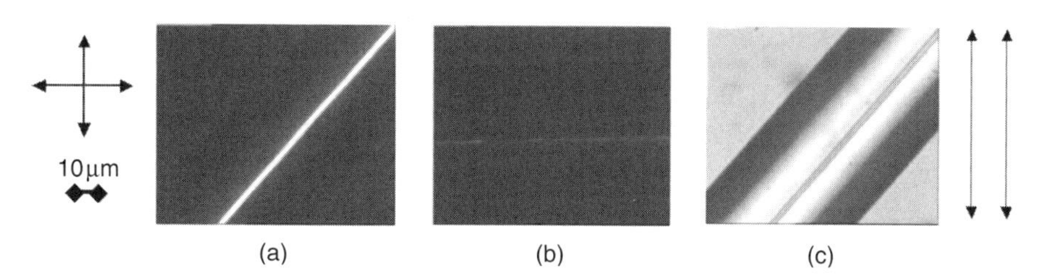

Figure 4.23 Photoalignment in microtube [67]. Crossed polarizers: (a) angle between polarizer and tube axis is 45°; (b) angle between polarizer and tube axis is 0°. Parallel polarizers: (c) angle between polarizer and tube axis is 45°. Reproduced from V. Presnyakov, Z. Liu, and V. Chigrinov, Infiltration of photonic crystal fiber with liquid crystals. *Proceedings of the SPIE* **6017**, 102 (2005), SPIE

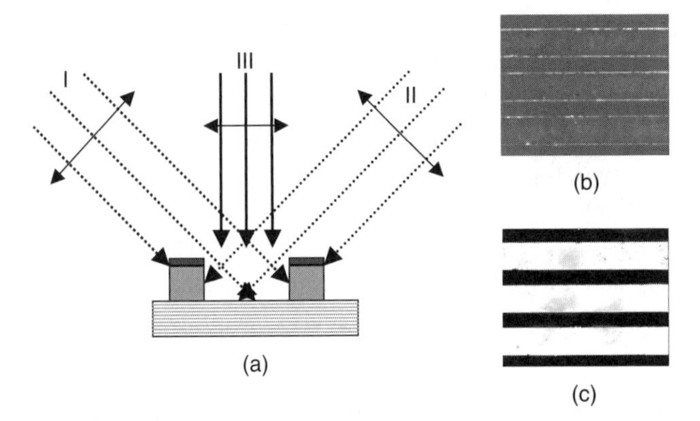

Figure 4.24 Optical scheme of three-step 3D substrate illumination: (a) profiled substrate LC cell between crossed (b) and parallel (c) polarizers [68]. Reproduced from A. Muravsky, V. Chigrinov, S. Studentsov, V. Brezhnev, N. Zhukov, and B. Gorfinkel, Photo-induced alignment technology for 3-D surface profiles of LCD substrates. *International Workshop on Liquid Crystals For Photonics, Gent, Belgium*, O-02 (2006)

The technique presented can be used as a non-contact method of LC alignment in complex photonic crystal structures.

Uniform azo-dye photoalignment on a profiled 3D surface (substrate with bulk relief) has been demonstrated [68] and a three-step exposure process for uniform surface alignment has been developed (Figure 4.24). Patterned exposure of azo-materials is a very useful procedure for LC photonic devices as a high-quality LC alignment on profiled surfaces can be obtained.

LC alignment on submicrometer-sized rib waveguides on silicon chips has also been studied [69]. The experiments, which used nematic LC cladding on silicon waveguides and microrings coated with a photoalignment layer and covered with a vertically aligned polyimide rubbed glass, revealed a defect-free hybrid aligned LCC on a silicon substrate (Figure 4.25). An electrically tunable microresonator using photoaligned LC as the cladding layer was demonstrated, where a photoalignment layer on the device surface defined the orientation of the LC molecules, and the transmission property of the waveguide-coupled microresonator was electrically tuned by varying the cladding refractive index under an applied electric field in the direction perpendicular to the microresonator plane (Figure 4.25) [69]. A tunable range of 20 nm was obtained for the microresonator, which is comparable with the thermo-optic effect usually used for the purpose.

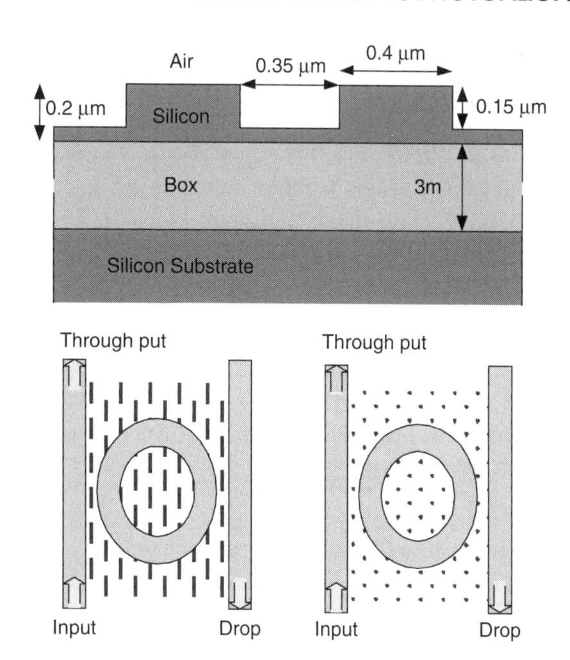

Figure 4.25 Schematic profile of submicrometer-sized rib Si microring system (upper) and LC orientation on photoaligned cladding azo-dye layer (lower) [69]. Reproduced from G. Chigrinov, L. Zhou, A. A. Muravsky, and A. W. Poon, Electrically tunable microresonators using photoaligned liquid crystals as cladding layers. US Provisional Patent Application No. 60/794,128, filed on April 24, 2006

The bulk-induced photoalignment of nematic LCs by photo-polymerization has been reported [70]. A nematic LC was mixed homogeneously with a small amount of the photoresist PVMC (Figure 2.13) and sandwiched between ITO-coated glass plates without any previous surface treatment. The sample was polymerized in its isotropic phase using linearly polarized UV light. On cooling it to the nematic state, a uniform orientation was obtained. The electro-optical properties of the volume photoaligned LC cells were almost the same as those prepared by rubbing. Homeotropic photoalignment of very high quality was produced when the photoresist was polymerized just a few degrees centigrade above the isotropic transition temperature of the nematic. The electro-optic characteristics of a volume photoaligned homeotropic LC cell containing PVMC were identical to those of standard homeotropic cells. As in planar cells, the photoresist does not alter the switching properties in any significant way.

A new method to form LC droplets in a polymer PVMC matrix has been proposed [71]. A planar orientation with a high order parameter $S = 0.7$ of the droplet optical axis is caused by light-induced anisotropy in the PVMC matrix. The proposed method has prospects for obtaining a stable orientation of LC droplets in polymer dispersed (PDLC) films without stretching the matrix.

References

[1] V. G. Chigrinov, *Liquid Crystal Devices: Physics and Applications*, Artech House, London (1999).

[2] S. Furumi, M. Nakagawa, Sh. Morino, K. Ichimura, and H. Ogasawara, Photogeneration of high pretilt angles of nematic liquid crystals by azobenzene-containing polymer films. *Applied Physics Letters* **74**, 2438 (1999).

[3] J.-Y. Hwang and D.-S. Seo, Control of high pretilt angle for nematic liquid crystal using *in situ* photoalignment method on homeotropic alignment layer. *Japanese Journal of Applied Physics* **40**, 4160 (2001).

[4] D.-S. Seo, D.-S. Park, and H.-J. Jeon, Response time mechanism for a photoaligned vertical-alignment liquid crystal display on a homeotropic alignment layer. *Liquid Crystals* **27**, 1189 (2000).

[5] J.-Y. Hwang, D.-S. Seo, J.-Y. Kim, and T.-H. Kim, Electro-optical characteristics in the photoaligned vertical-alignment cell on a photopolymer surfaces containing N-(phenyl)maleimide. *Japanese Journal of Applied Physics* **41**, L58 (2002).

[6] S. C. Jain, V. K. Tanwar, and V. Dix, Homeotropic and planar orientation of liquid crystals on a photopolymer coated surface irradiated with UV-light. *Japanese Journal of Applied Physics* **41**, L1106 (2002).

[7] M. Schadt, Photoaligned liquid crystal displays and LC-polymer optical films. *Euro Display'99 Digest*, p. 27 (1999).

[8] H. Seiberle, K. Schmitt, and M. Schadt, Multidomain LCDs and complex optical retarders generated by photoalignment. *Euro Display'99 Digest*, p. 121 (1999).

[9] J.-H. Park, Y. Choi, T.-Y. Yoon, C.-J. Yu, and S.-D. Lee, A self-aligned multi-domain liquid-crystal display on polymer gratings in a vertically aligned configuration. *Journal of the SID* **11**/2, 283 (2003).

[10] V. Chigrinov, H.-S. Kwok, H. Takada, and H. Takatsu, Photoaligning by azo-dyes: physics and applications. *Liquid Crystals Today* **4**, 1 (2005).

[11] V. A. Konovalov, V. G. Chigrinov, H.-S. Kwok, H. Takada, and H. Takatsu, Photoaligned vertical aligned nematic mode in liquid crystals. *Japanese Journal of Applied Physics* **43**, 261 (2004).

[12] J. S. Lee, K. Y. Han, B. H. Chae, G. B. Park, and W. S. Park, Wide viewing angle 4-domain HTN (Hybrid Twisted Nematic) LCD by using photoalignment method to apply TFT LCD. *Asia Display'98 Digest*, p. 781 (1998).

[13] J. Chen, P. J. Bos, D. R. Bryant, D. L. Johnson, S. H. Jamal, and J. R. Kelly, Simple four-domain twisted nematic liquid crystal display. *Applied Physics Letters* **67**, 1990 (1995).
[14] N. A. Aerle, A novel multi-domain wide-viewing angle liquid crystal display. *Japanese Journal of Applied Physics* **34**, L1472 (1995).
[15] S.-D. Cheng and Z.-M. Sun, Amorphous twisted nematic liquid crystal displays: the characteristics and theoretical considerations. *Liquid Crystals* **19**, 321 (1995).
[16] M. Schadt, H. Seiberle, and A. Schuster, Optical patterning of multi-domain liquid-crystal displays with wide viewing angles. *Letters to Nature* **381**, 212 (1996).
[17] V. A. Konovalov, A. A. Minko, A. A. Muravski, S. N. Timofeev, and S. E. Yakovenko, Mechanism and electro-optic properties of multidomain vertically aligned mode. *Journal of the SID* **7**/3, 213 (1999).
[18] C. Nieuwkerk, T. Mol, S. Stallinga, F. Leenhouts, P. van de Witte, J. van Haaren, and D. Broer, Economic wide-viewing angle AMLCD using dual-domain photoalignment and a single discotic compensation foil. *SID'00 Digest*, p. 842 (2000).
[19] D. K. G. de Boer, A. C. Nieuwkerk, R. Cortie, F. A. Fernández, S. E. Day, and R. James, Domain walls in photoaligned double-domain twisted-nematic LCDs. *Journal of the SID* **11**/3, 443 (2003).
[20] D. D. Huang, V. Kozenkov, Vl. Chigrinov, H.-S. Kwok, H. Takada, and H. Takatsu, Novel photoaligned nematic liquid crystal cell. *Japanese Journal of Applied Physics* **44**, 5117 (2005).
[21] M. Kimura, S. Nakata, Y. Makita, Y. Mastuki, A. Kumano, Y. Takeuchi, and H. Yokoyama, New photoalignment technology for IPS-LCDs. *SID'01 Digest*, p. 1162 (2001).
[22] V. Vorflusev, V. Kosenkov, and V. Chigrinov, Bistable switching in FLC cells aligned by photoanisotropic films. *Molecular Crystals and Liquid Crystals* **263**, 577 (1995).
[23] J. Funfschilling, M. Stalder, and M. Schadt, Alignment of ferroelectric LCDS with hybrid LPP-photoaligned polymer networks. *SID'99 Digest*, p. 308 (1999).
[24] W.-S. Kang, H.-W. Kim, and J.-D. Kim, Zigzag defect-free alignment of surface stabilized ferroelectric liquid crystal cells with a polyimide irradiated by polarized UV light. *Liquid Crystals* **28**, 1715 (2001).
[25] W.-S. Kang, H.-W. Kim, and J.-D. Kim, Contrast ratio and switching of zigzag defect-free surface atabilized FLCD by photoinduced alignment. *Liquid Crystals* **29**, 583 (2002).
[26] H. Furue, Y. Iimura, Y. Miyamoto, H. Endoh, H. Fukuro, and Sh. Kobayashi, Fabrication of a zigzag defect-free surface-stabilized ferroelectric liquid crystal display using polyimide orientation film. *Japanese Journal of Applied Physics* **37**, 3417 (1998).
[27] R. Kurihara, H. Furue, T. Takahashi, T. Yamashita, J. Xu, and Sh. Kobayashi, Fabrication of defect-free ferroelectric liquid crystal displays using photoalignment and their electrooptic performance. *Japanese Journal of Applied Physics* **40**, 4622 (2001).
[28] J. Xu, R. Kurihara, and Sh. Kobayashi, Phenomenological analysis in fabricating defect-free ferroelectric liquid crystal displays with C-1 or C-2 uniform states. *Japanese Journal of Applied Physics* **40**, 4626 (2001).

[29] T. Ishitani, H. Furuta, J. Xu, and Sh. Kobayashi, Influence of the surface alignment conditions and the polymer stabilization on the electrooptic characteristics of ferroelectric liquid crystal displays exhibiting half-V switching – a comparison of photoalignment and rubbing technique. *Japanese Journal of Applied Physics* **40**, L973 (2001).

[30] T. Fujisawa, M. Hayashi, H. Hasebe, K. Takeuchi, H. Takatsu, and Sh. Kobayashi, Novel PSV-FLCDs with high response speed, high optical throughput, and high contrast ratio with small voltage shift by temperature: application to field sequential full color LCDs. *SID'07 Digest*, p. 633 (2007).

[31] D. D. Huang, E. P. Pozhidaev, V. G. Chigrinov, H. L. Cheung, Y. L. Ho, and H.-S. Kwok, Photoaligned ferroelectric liquid crystal displays based on azo-dye layers. *Displays* **25**, 21 (2004).

[32] E. Pozhidaev, V. Chigrinov, and X. Li, Photoaligned ferroelectric liquid crystal passive matrix display with memorized gray scale. *Japanese Journal of Applied Physics* **45**, 875 (2006).

[33] E. P. Pozhidaev and V. G. Chigrinov, Bistable and multistable states in ferroelectric liquid crystals. *Crystal Reports* **51**, 1030 (2006).

[34] S. Valyukh, I. Valyukh, P. Xu, and V. Chigrinov, Study on birefringent color generation for a reflective ferroelectric liquid crystal display. *Japanese Journal of Applied Physics* **45**, 7819 (2006).

[35] P. Xu, X. Li, and V. Chigrinov, Double cells achromatic ferroelectric liquid crystal displays using photoalignment technology. *Japanese Journal of Applied Physics* **45**, 200 (2006).

[36] S.-Y. Hyun, J.-H. Moon, and D.-M. Shin, Application of photoalignment technologies to antiferroelectric liquid crystal cell. *Molecular Crystals and Liquid Crystals* **412**, 369 (2004).

[37] R. Yamaguchi, Y. Goto, and S. Sato, A novel patterning method of liquid crystal alignment by azimuthal anchoring control. *Japanese Journal of Applied Physics* **41**, L889 (2002)

[38] R. Yamaguchi and S. Sato, Continuous grey scale image printing on the liquid crystal cell. *Applied Physics Letters* **86**, 031913 (2005).

[39] R. Yamaguchi, Y. Goto, and S. Sato, Liquid crystal alignment surface with two easy axes induced by unidirectional rubbing. *Applied Physics Letters* **82**, 4450 (2003).

[40] R. Yamaguchi, T. Kawamura, and S. Sato, Dual image writing on the LC cell using unpolarized UV light. *IDW'04 Digest*, p. 39 (2004).

[41] R. Yamaguchi and S. Sato, Liquid crystal optical security device with polarized latent images. *IDW'05 Digest*, p. 65 (2005).

[42] T.-H. Lin, H.-C. Jau, S.-Y. Hung, H.-R. Fuh, and A. Y.-G. Fuh, Photoaddressable reflective liquid crystal display. *Applied Physics Letters* **89**, 021116 (2006).

[43] L. T. Thieghi, J. J. Bonvent E. A. Oliveira, J. A. Giacometti, and D. T. Balogh, Competition between anchoring and reversible photo-induced alignment of a nematic liquid crystal. *Applied Physics A* **77**, 911 (2003).

[44] Al. Muravsky, An. Murauski, X. Li, V. Chigrinov, and H.-S. Kwok, Optical rewritable liquid-crystal-alignment technology. *Journal of the SID* **15**/4, 267 (2007).

[45] P. Martinot-Lagarde and I. Dozov, The Binem display, a nematic bistable device switched by surface anchoring breaking. *Proceedings of the SPIE* **5003**, 25 (2003).
[46] F. Yeung and H.-S. Kwok, Truly bistable twisted nematic liquid crystal display using photoalignment technology. *Applied Physics Letters* **83**, 4291 (2003).
[47] X. Li, F. Yeung, V. G. Chigrinov, and H.-S. Kwok, Flexible photoaligned permanent bistable TN-LCD. *IDW'05 Digest*, p. 883 (2005).
[48] G. P. Crawford (Ed.), *Flexible Flat Panel Displays*, John Wiley & Sons, Ltd, Chichester (2005).
[49] J.-Y. Hwang, K.-H. Nam, and D.-S. Seo, Alignment properties and EO performances of flexible TN-LCD using in-situ photoalignment method with the polymer film. *Molecular Crystals and Liquid Crystals* **434**, 143 (2005).
[50] J. Osterman, A. Tong, K. Skarp, V. Chigrinov, and H.-S. Kwok, Properties of azo-dye alignment on plastic substrates. *Journal of the SID* **13**, 1003 (2005).
[51] K. Maruyama, Y. Ono, Y. Suzuki, and T. Ikariya, A plastic color TN-LCD using a photo alignment method. *IDW'05 Digest*, p. 141 (2005).
[52] X. Tong, A. Natansohn, and P. Rochon, Photoinduced liquid crystal alignment based on a surface relief grating in an assembled cell. *Applied Physics Letters* **74**, 3791 (1999).
[53] A. Parfenov, N. Tamaoki, and S. Ohnishi, Photoinduced alignment of nematic liquid crystal on the polymer surface microrelief. *Journal of Applied Physics* **87**, 2043 (2000).
[54] D. Dantsker, J. Kumar, and S. K. Tripathy, Optical alignment of liquid crystals. *Journal of Applied Physics* **89**, 4318 (2001).
[55] S. S. Bartkiewicz, K. Matczyszyn, J. Mysliwiec, O. Yaroshchuk, T. Kosa, and P. Palffy-Muharey, LC alignment controlled by photoordering and photorefraction in a command surface. *Molecular Crystals and Liquid Crystals* **412**, 301 (2004).
[56] X. Lu, Q. Lu, and Z. Zhu, Alignment mechanism of a nematic liquid crystal on a prerubbed polyimide film with laser-induced periodic surface structure. *Liquid Crystals* **30**, 985 (2003).
[57] V. Presnyakov, K. Asatryan, T. Galstian, and V. Chigrinov, Optical polarization grating induced liquid crystal micro-structure using azo-dye command layer. *Optics Express* **14**, 10558 (2006).
[58] C. Oh and M. J. Escuti, Achromatic diffraction using reactive mesogen polarization gratings. *SID'07 Digest*, p. 1401 (2007).
[59] G. P. Crawford, J. N. Eakin, M. D. Radcliffe, A. Callan-Jones, and R. A. Pelcovits, Liquid-crystal diffraction gratings using polarization holography alignment techniques. *Journal of Applied Physics* **98**, 123102 (2005).
[60] Y. Ukai, T. Ohyama, L. Fennell, Y. Kato, M. Paukshto, P. Smith, O. Yamashita, and S. Nakanishi, Current status and future prospect of in-cell-polarizer technology. *Journal of the SID* **13**/1, 17 (2005).
[61] V. M. Kozenkov, W. C. Yip, S. T. Tang, V. G. Chigrinov, and H.-S. Kwok, Thin photo-patterned internal polarizers for LCDs. *SID'00 Digest*, p. 1099 (2000).
[62] W. C. Yip, H.-S. Kwok, V. M. Kozenkov, and V. G. Chigrinov, Photo-patterned e-wave polarizer. *Displays* **22**, 27 (2001).

[63] K. Ichimura, T. Fujiwara, M. Momose, and D. Matsunaga, Surface-assisted photoalignment control of lyotropic liquid crystals, Part 1. Characterisation and photoalignment of aqueous solutions of a water-soluble dye as lyotropic liquid crystals. *Journal of Materials Chemistry* **12**, 3380 (2002).

[64] K. Ichimura, S. Furumi, S. Morino, M. Kidowaki, M. Nakagawa, M. Ogawa, and Y. Nishiura, Photoinduced orientation of discotic liquid crystals. *Advanced Materials* **12**, 950 (2000).

[65] S. Furumi, K. Ichimura, H. Sata, and Y. Nishiura, Photochemical manipulation of discotic liquid crystal alignment by a poly (vinyl cinnamate) thin film. *Applied Physics Letters* **17**, 2689 (2000).

[66] S. Furumi, M. Kidowaki, M. Ogawa, Y. Nishiura, and L. Ichimura, Surface-mediated photoalignment of discotic liquid crystals on azobenzene polymer films. *Journal of Chemical Physics* **109**, 9245 (2005).

[67] V. Presnyakov, Z. Liu, and V. Chigrinov, Infiltration of photonic crystal fiber with liquid crystals. *Proceedings of the SPIE* **6017**, 102 (2005).

[68] A. Muravsky, V. Chigrinov, S. Studentsov, V. Brezhnev, N. Zhukov, and B. Gorfinkel, Photo-induced alignment technology for 3-D surface profiles of LCD substrates. *International Workshop on Liquid Crystals for Photonics, Gent, Belgium*, O-02 (2006).

[69] G. Chigrinov, L. Zhou, A. A. Muravsky, and A. W. Poon, Electrically tunable microresonators using photoaligned liquid crystals as cladding layers. US Provisional Patent Application No. 60/794,128, filed on April 24, 2006.

[70] S. C. Jain and H. S. Kitzerow, Bulk-induced alignment of nematic liquid crystals by photopolymerization. *Applied Physics Letters* **64**, 2946 (1994).

[71] V. Nazarenko, Yu. Reznikov, V. Reshetnyak, V. Sergan, and V. Zyryanov, Oriented dispersion of liquid crystal droplets in a polymer matrix with light induced anisotropy. *Molecular Materials* **2**, 295 (1993).

5

Application of Photoalignment Materials in Optical Elements

5.1 Polarizers

5.1.1 Dichroism

Polarizers are widely used in optics, particularly in LCDs. Broadly speaking, there are two kinds of polarizers – absorptive and reflective. The former are based on anisotropic absorption of light along orthogonal directions on the polarizer film. The latter are based on multilayer optical coatings with an oblique angle of incidence, such as in polarizing beam splitters [1] or multilayers of anisotropic polymers [2]. These polarizers are lossless and present an advantage in many display applications, such as in projectors or in LCD backlights for maximum optical efficiency. However, in this discussion, we shall only concentrate on the former types, which are based on anisotropic absorbers [3].

Photoalignment of Liquid Crystalline Materials: Physics and Applications
V. Chigrinov, V. Kozenkov and H.-S. Kwok

Quantitatively, the transmittance of light along two orthogonal polarizations through an absorptive polarizer is given by Beer's law

$$T_{||} = \exp(-\alpha_{||} d), \quad T_{\perp} = \exp(-\alpha_{\perp} d). \tag{5.1}$$

Here α is the absorption coefficient and d is the absorptive film thickness. $T_{||}/T_{\perp}$ can be defined as the extinction ratio ε and $T_{\perp}$ can be defined as the transmission efficiency η. Of course, for a good polarizer ε should be very small and η should be as close to 100% as possible.

We define the dichroic ratio of the absorptive medium as

$$D = \frac{\alpha_{||}}{\alpha_{\perp}}. \tag{5.2}$$

Then it is easy to show that

$$\varepsilon = \eta^{D-1}. \tag{5.3}$$

Sometimes it is useful to define the contrast ratio of a polarizer as the inverse of the extinction ratio:

$$CR = \frac{1}{\varepsilon} \tag{5.4}$$

Figure 5.1 shows a plot of *CR* and the transmittance of the polarizer for various values of D. For unpolarized light input, the transmittance of the polarizer is 0.5 η. From the above expressions, it is obvious that in order to maximize both *CR* and η, the dichroic ratio of the absorptive medium should be as large as possible. For

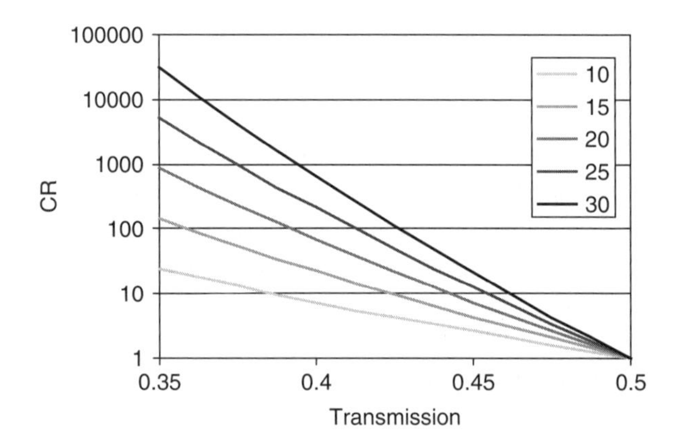

Figure 5.1 Plot of the contrast ratio *CR* and transmittance η of the polarizer for various values of dichroic ratio D

example, for a CR of 1000 and η of 0.9, the required dichroic ratio is $D = 66.5$. If $CR = 1000$ and $D = 20$, η is only 69.5%, which is rather low.

Thus for reasonable performance polarizers, the dichroic ratio has to be over 25. Current polarizers are based on iodine impregnation of stretched polymer films [4] (e.g. 30–50 μm polyvinyl alcohol). The dichroic ratio is over 50. There are two issues in getting high performance from polarizers based on anisotropic absorption. First, the intrinsic absorption anisotropy has to be large. Second, the absorber has to be well aligned in one direction. Both are difficult issues.

A standard method to characterize the quality of a polarizer is to measure the transmittance of unpolarized light through two polarizers with axes in parallel and perpendicular respectively. It is easy to show that the measured values are related to the absorption coefficients by

$$\overline{T}_{||} = 0.5[\exp(-2\alpha_{||} d) + \exp(-2\alpha_{\perp} d)], \quad \overline{T}_{\perp} = \exp[-(\alpha_{||} + \alpha_{\perp})]. \tag{5.5}$$

It is straightforward to see that such a measurement can yield the dichroic ratio of the absorber as well as its spectral behavior. Figure 5.2 shows such a measurement for a commercial polarizer.

From these curves, it can be seen that the peak optical efficiency is about 91% while the contrast ratio is 200. The implied dichroic ratio is 60. In the literature, values for the dichroic ratio of 50 or higher have been reported [5]. Also, as is

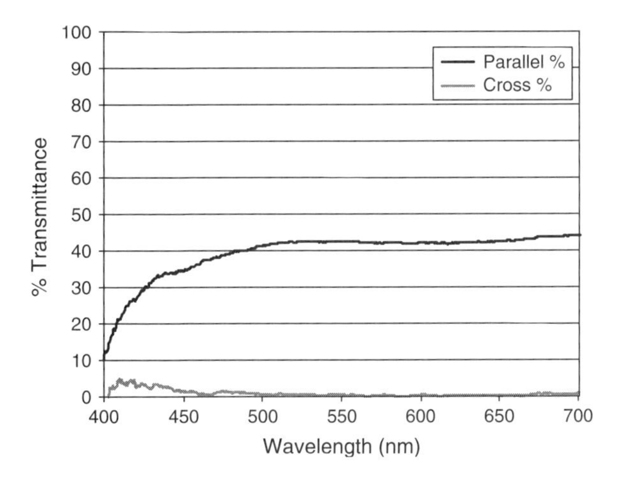

Figure 5.2 Wavelength dependence of the transmission of a commercial polarizer

typical for iodine-based commercial polarizers, there is some light leakage in the blue region of the spectrum.

Anisotropic dye materials are abundant. But the dichroic ratio as well as the spectral coverage are still insufficient for making good polarizers to cover the entire visible spectrum. However, this is mostly a materials science and chemistry issue and should be solvable. There are indeed good materials such as iodine that are good anisotropic absorbers. A combination of several dyes can also overcome the spectral coverage problem. The issue of alignment is the one that will be addressed here. In existing technologies, alignment of the anisotropic absorber is achieved by stretching a polymeric film. Uniformity of stretching is important to obtain uniform dichroism. Stretching the length by five times in one direction is not unusual to obtain good directionality. The technology of achieving uniform stretching is not simple, thus the manufacture of polarizers is challenging.

There are other ways of aligning a dichroic absorber. Mechanical and photoalignment have been explored and are successful to some extent. In mechanical alignment, the liquid film of the absorbing material is brushed in one direction. Provided good brushing is employed, alignment can be achieved quite well [6]. Lyotropic LCs which absorb in the visible can be used for such purposes. Such a 'liquid polarizer' has a dichroic ratio of up to 20 and can be useful in low-end LCDs. The alignment is achieved through mechanical shearing of the lyotropic LC. However, the problem of large dichroic ratio and uniform alignment is still an issue with such mechanical methods.

Photoalignment is an alternative to mechanical rubbing. It has many advantages, the most important of which is the high degree of uniformity that can be achieved in principle. Another one is that a good order parameter can be obtained, which will increase the effective dichroic ratio of the absorber. There are several possibilities for photoalignment of polarizers. One way is to directly align the dichroic absorber itself. The other way is to photoalign an intermediate layer that will align the dichroic dye. Yet another way is to align a polymer and use that as a host to align the dichroic absorber. The ultimate goal is to induce a high order parameter of the absorbing material.

5.1.2 Direct Photoalignment

Figure 5.3 shows an azo-dye which can be photoaligned directly [7]. The material is azo-dye AD-1. This dichroic dye is an azo-dye that can be aligned by UV radiation by the mechanism of orientational diffusion.

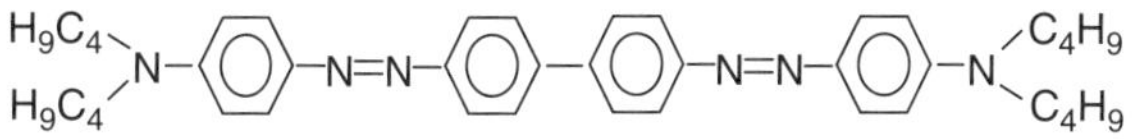

H_9C_4, H_9C_4 N—⟨⟩—N=N—⟨⟩—⟨⟩—N=N—⟨⟩—N C_4H_9, C_4H_9

Figure 5.3 Photoaligned dichroic azo-dye AD-1 [7]. Reproduced from W. C. Yip, H.-S. Kwok, V. M. Kozenkov, and V. G. Chigrinov, Photo-patterned e-wave polarizer. *Displays* **22**, 27 (2001), Elsevier

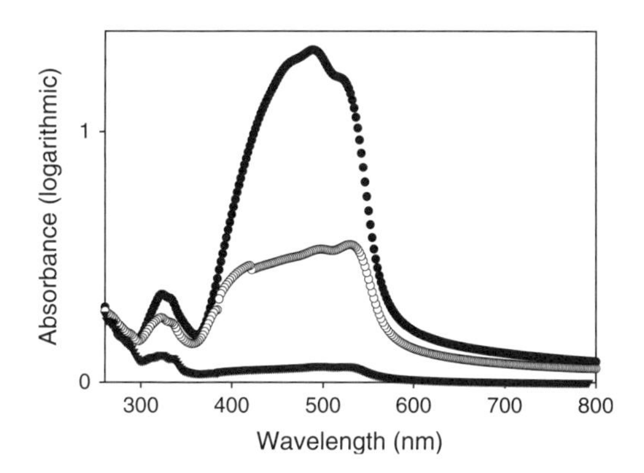

Figure 5.4 The dichroic absorption spectra of the AD-1 dye after and before UV photoalignment. The upper and lower curves are the absorption spectra after irradiation along two orthogonal directions. The middle one is before irradiation

The dosage and anchoring energy results are similar to those of sulfonic dye SD1 (Figure 2.4), another standard photoalignable azo-dye. The dichroic absorption spectra of the dye after and before UV photoalignment are shown in Figure 5.4. In this figure, the upper and lower curves are the absorption spectra after irradiation along two orthogonal directions. The middle one is before irradiation.

It can be seen from Figure 5.4 that photoalignment can induce dichroic absorption in AD-1. The dichroic ratio can reach a value of about 20, which is not quite good enough yet for practical applications. Further work is probably needed to improve the dichroic ratio. As discussed below, this dichroic ratio can be improved to 25 if proper treatment is used.

It can also be seen from Figure 5.4 that the absorption spectrum does not cover the entire visible spectrum. Hence another dye or more dyes are needed to cover the blue and red regions. Presumably a mixture of dyes similar to AD-1 can be used.

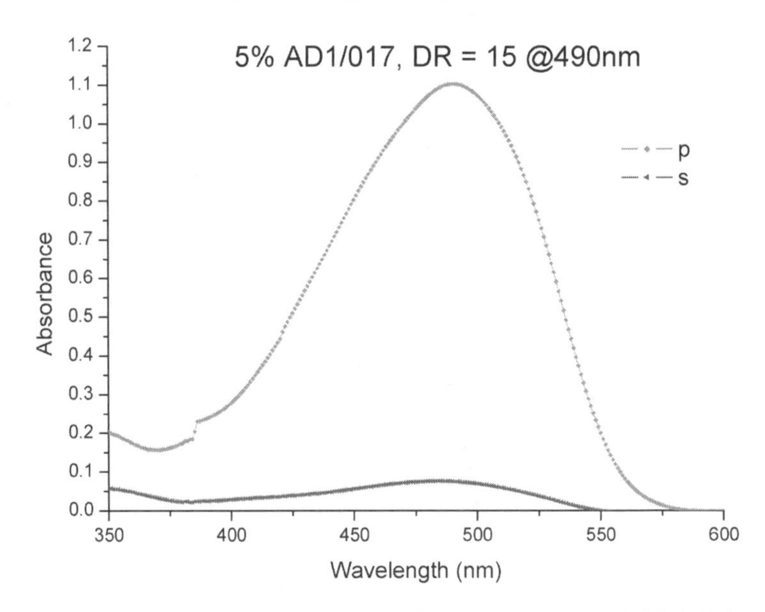

Figure 5.5 Dichroic spectra of AD-1 mixed with a polymer agent LCP 017 from DIC (Japan); p- and s-polarizations of the light correspond to the directions parallel and perpendicular to the preferred photoaligned directions accordingly

As in other azo-dyes, the film needs to be stabilized after photoalignment. This can be accomplished as usual by adding an appropriate amount of polymer to the film. Figure 5.5 shows a film of AD-1 mixed with a polymer agent. Dichroism is still maintained after polymerization. The film is stable against strong light irradiation, rendering it more practical and suitable for real applications. However, the dichroic ratio is reduced somewhat by the polymer.

In a recent experiment, it was found that the order parameter of the dichroic dye can be improved greatly if the photoalignment layer is applied by vapor phase deposition of the AD-1 dye [8]. For example, using AD-1, a dichroic ratio of 35 can be achieved. This is larger than the ratio of 20 obtained by photoalignment of a spin-coated AD-1 film. Thus, photoalignment becomes extra-effective in achieving a large order parameter. An *in situ* alignment method with vapor deposition during photoalignment can also be used to provide a high order parameter with low exposure energy [8].

5.1.3 Indirect Photoalignment

For an anisotropic absorber that cannot be directly photoaligned, an indirect method is needed. As mentioned previously, many lyotropic discotic LCs can be good dichroic absorbers that can be aligned by a shearing force. They are e-wave polarizers since the *c* axis lies along the plane of absorption and they have unique optical properties that are different from conventional o-wave polarizers when an oblique angle of incidence is involved [6]. The discotic molecules tend to stack up. Hence a mechanical shearing force will align them in the liquid state. Upon drying an anisotropic absorber is obtained.

While mechanical shearing to align such e-wave polarizers can be achieved easily, the issue of uniformity and manufacturability is unclear. Alternative methods of photoalignment offer some special features such as patternability [7]. It turns out that lyotropic LCs can be aligned by another alignment layer. While ordinary rubbed polyimide can align the polarizer liquid, photoalignment can be used as well. Figure 5.6 shows the anisotropic absorption spectra of a lyotropic LC on AD-1 which is photoaligned by UV light similar to the case of Section 5.1.2.

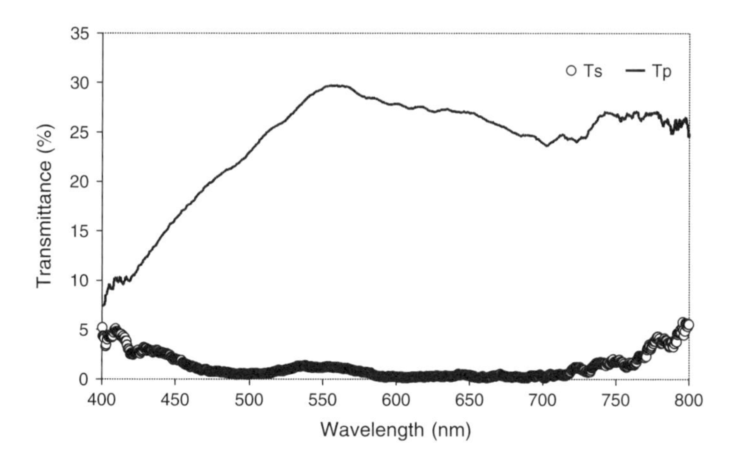

Figure 5.6 Polarized transmission spectra of a photoaligned lyotropic LC polarizer. T_p and T_s represent the transmission spectra parallel and perpendicular to the direction of the polarized UV light respectively [7]. Reproduced from W. C. Yip, H.-S. Kwok, V. M. Kozenkov, and V. G. Chigrinov, Photo-patterned e-wave polarizer. *Displays* **22**, 27 (2001), Elsevier

The dichroic ratio achievable depends on the lyotropic LC itself. For the data shown in Figure 5.6, the dichroic ratio is 13, which is not quite large enough. This value is actually slightly larger than that obtained by mechanical shearing. This is because the order parameter is probably larger in the case of alignment on the molecular level.

5.1.4 Patterned Polarizers

The application of photoalignment to the fabrication of polarizers has a unique advantage over other methods such as mechanical shearing. The polarizer can be patterned – both the spatial pattern as well as the orientation of the optical axis can be finely patterned by photolithography. This may offer a new technique for making many new displays, such as a 3D display with the left and right eye seeing different polarizations.

Figure 5.7 shows a patterned polarizer [7]. In this case the polarizer has a fixed pattern. Presumably it can also be a pixelated pattern. The color of the polarizer

Figure 5.7 The photoinduced pattern of the HKUST logo viewed with a polarizer with a polarization axis perpendicular to the absorption axis of the azo-dye AD-1 [7]. Reproduced from W. C. Yip, H.-S. Kwok, V. M. Kozenkov, and V. G. Chigrinov, Photo-patterned e-wave polarizer. *Displays* **22**, 27 (2001), Elsevier

is due to the incomplete absorption spectrum of the dichroic material across the entire visible spectrum; in this case it is AD-1.

It is more useful to pattern the polarizer in pixelated form. For example, it may be possible to make transflective displays with subpixels having different polarizer orientations, with one subpixel being in the transmissive mode and the other one being in the reflective mode. We shall discuss such a possibility in Section 5.3.

Another application of pixelated polarizers is in making 3D displays. It is possible to have the left and right eye seeing different pictures on the same LCD screen if alternate pixels have different polarized output. Thus a simple eye glass with complementary polarizers will allow a 3D image to be observed.

5.2 Retardation Films

5.2.1 Types of Films

Photoalignment can be applied to the fabrication of retardation films as well. Retardation films are used widely in LCDs for viewing angle enhancement as well as dispersion compensation. They can be made from uniaxial or biaxial materials. Within the class of uniaxial materials, many types of films can be made and they serve different functions. For example, for uniform films where the optical axis is fixed throughout the entire film, one can have A-plates, C-plates and O-plates as shown in Figure 5.8.

They correspond to the homogeneous, homeotropic, and tilted alignments using the language of LC optics. The birefringence of the uniaxial material can be positive ($n_e > n_o$) or negative ($n_e < n_o$). Thus one can have positive A-plates and negative A-plates, and similarly for the other types.

While positive A-plates are most common, negative C-plates are gaining popularity, in particular for compensation of the vertically aligned nematic (VAN) mode. Take for example the simple VAN with the LC cell rubbed at 45° to the crossed input and output polarizers. At normal incidence, the non-select state is totally dark giving excellent contrast ratio for this mode. However, at oblique angles, as seen in Figure 5.9, there is a retardation between the ordinary and extraordinary waves.

The extraordinary index $n_e(\theta)$ is given by

$$\frac{1}{n_e^2(\theta)} = \frac{\cos^2\theta}{n_o^2} + \frac{\sin^2\theta}{n_e^2} \tag{5.6}$$

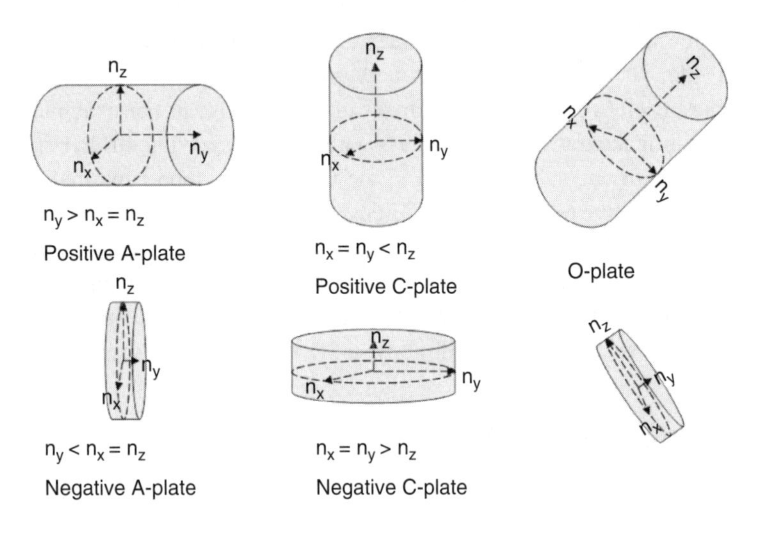

Figure 5.8 Various possibilities of phase retarders

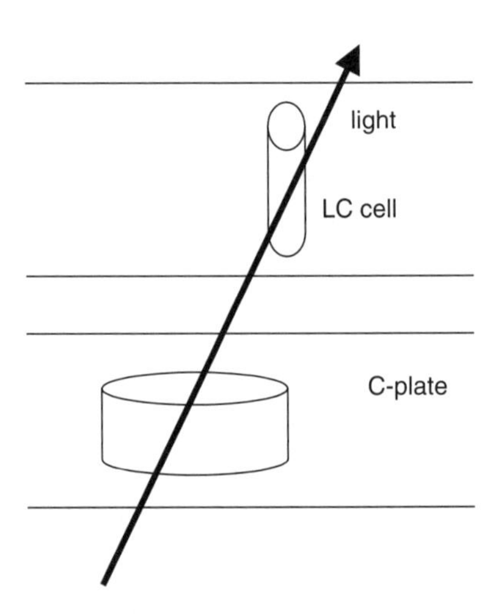

Figure 5.9 Compensation of VAN LC configuration by a negative C-plate

where θ is the angle of incidence of the light. For light going through the LC cell, $n_e(\theta)$ is larger than n_o. The net retardation of the light going through the LC cell is $d(n_e(\theta) - n_o)$ where d is the cell thickness. Thus this will lead to light leakage and loss of contrast. However, if a negative C-plate is placed on the LC cell, then the same light will also experience a net retardation through the retardation film with a value of $d'(n'_e(q) - n'_o)$ where the prime indicates the value for the C-plate. This net retardation of light going through the C-plate is negative. Thus if

$$d(n_e(\theta) - n_o) + d'(n'_e(\theta) - n'_o) = 0 \tag{5.7}$$

then the total net retardation between the o-wave and the e-wave will be zero at any angle of incidence. The vertically aligned cell will be totally dark for all values of θ. Note that because of uniaxial symmetry, there is no dependence on the azimuthal angle ϕ and this compensation is completely satisfied for all angles of incidence. The only complication with this method is that we have to find a negative C-plate with a birefringence that has the same wavelength dispersion and temperature dependence as the LC material. Also in this simple illustration, we have not considered the leakage of light through the crossed polarizers [9]. That problem can be taken care of by using broadband quarter-wave plates [10].

In addition to retardation films where the optical axis is uniform, it is now possible to make retardation films with a variable optical axis. This is possible with developments in the synthesis of polymerizable LC materials or LC polymers. The photoalignment of all types of films will be discussed next.

5.2.2 Direct and Indirect Photoalignment

Retardation films are thus extremely useful in making LCD with wide viewing angles. As in the case of polarizers, the uniaxial materials forming the retardation films can be aligned either directly or indirectly. In the direct alignment method, a photoalignable material has to be the birefringent material itself. This is indeed similar to the case of photoalignment of an alignment layer. Since a retardation film has to be rather thick, low optical absorption of the material is important for the film to be transparent. Obviously this method has a major limitation in that the material requirement is too demanding. Most photoalignable materials are somewhat absorptive in the visible. For example, it is possible to photoalign SD1 layer by layer up to a large thickness to form a retardation film. But nearly a micrometer of the film is needed, rendering the film rather yellow. Likewise, photoaligned AD-1 is a polarizer rather than a retarder.

In principle, it is possible to make use of the guest–host effect to fabricate a retardation film. That is, one can mix the photoalignable material (guest) together with another polymerizable birefringent material (host). Photoalignment of the guest will align the host as well. Rapid UV polymerization of the host will then produce a retardation film. There are two ways that this can be accomplished: (i) use a crosslinkable photopolymer; or (ii) the photoalignable mesogen is part of the polymer chain. The photoactive mesogen interacts directly with the controlling light and can be oriented to form a retardation film. Morino *et al.* described a binary mixture of nematic LC and a crosslinkable photopolymer [11]. A retardation film can be produced with linearly polarized light irradiation.

However, there are problems with this direct approach. The method is quite demanding on the material, which has to be photoalignable, should be easily incorporated into the polymer chain, and should be transparent. Also, if the film is too thick, alignment of the entire film becomes difficult as the UV light is absorbed near the surface. In addition, it is necessary for the photoaligning light to have a different wavelength from the UV curing light. The photoaligning light should not polymerize the host. Thus the direct method is difficult and by and large not successful in producing good-quality retardation films.

An alternative method is to separate the photoalignment requirement and the polymerizability requirement. The retardation material can be aligned by an alignment layer or layers. This is possible if the retardation material has LC properties. Since the retardation material is in most cases a polymer, this type of material is therefore exactly a liquid crystal polymer (LCP) [12]. LCP is also sometimes known as polymerizable liquid crystal (PLC). It was studied for a long time for many other applications before the possibility of fabricating retardation films was considered. LCPs are mechanically very strong and are used in the production of very reliable plastic materials, e.g. in making bullet-proof vests. Indeed, the synthesis of LCP with good mechanical and optical properties is a major area of research in polymer chemistry.

LCPs monomers naturally align themselves because of their LC properties. This alignment is the main reason for the strong mechanical properties of LCPs. However, such self-assembly necessarily will contain domains. For optical applications, uniform alignment and uniform optical properties are needed. Thus, the LCP should be aligned by an alignment layer. This alignment can be in the form of mechanical rubbing or photoinduced alignment, just like conventional LCs.

LCP can be classified into two main types: main-chain LCP or side-chain LCP. The situation is illustrated in Figure 5.10.

The LCP consists of an LC mesogen and a polymeric chain. The mesogen can be incorporated into the main chain of the polymer or appears as a side chain – hence the terms main-chain LCP and side-chain LCP. For the case of side chains, the LC mesogen can be grafted perpendicularly or parallel to the main chain as shown

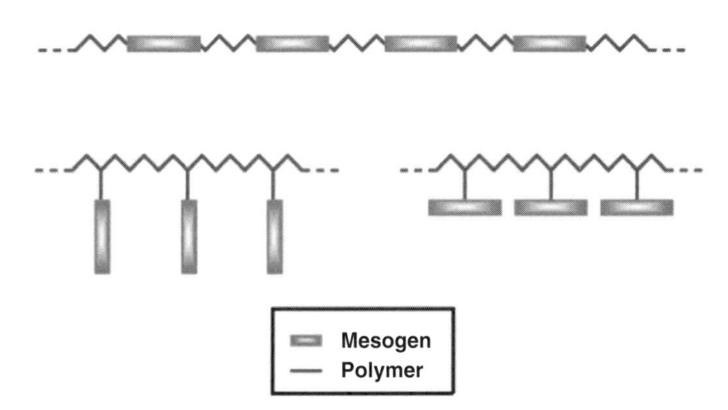

Figure 5.10 The structure of LCP film: main-chain LCP (upper) and side-chain LCP (lower) [12]. Reproduced from H. Hasebe, K. Takeuchi, and H. Takatsu, Properties of UV curable liquid crystals and their retardation. *Journal of the SID* **3**, 139 (1995), The Society for Information Display

in Figure 5.10. Obviously there are differences in mechanical as well as optical properties for these different types of LCP. The birefringence can come from either the LC mesogen or the polymer itself, or both.

The alignment of LCPs is an interesting topic. Obviously, for optical applications, the LCP should be transparent in the visible, should have a large birefringence, can be polymerized rapidly with a small light dosage, can be aligned well (good order parameter) in the liquid state, and should be stable against UV radiation. The point about rapid photo-polymerization is important. After the polymer precursor is aligned, either mechanically or optically, it has to be polymerized. The polymerization process will destroy the original alignment if it is not carried out rapidly. It is well known that polymers shrink or expands upon polymerization. Thus it is important to lock the polymer in place during the process of polymerization.

5.2.3 Examples of Photoaligned Retardation Films

Figure 5.11 shows an example of an LCP film that has been aligned by a photoalignment layer SD1. The material is RMM256C (E. Merck), which is a commercially available reactive mesogen. The figure shows the photoaligned film wedged between two crossed polarizers at either 0° or 45° to the polarizers. It can

Figure 5.11 Photographs of anisotropic RMM256C film viewed between a pair of crossed polarizers. The film is uniformly aligned by an SD1 photoalignment layer in the long-side direction of the rectangular glass slide (slow axis of film) [13]. Reproduced from Oleg Yaroshchuk, Jacob Ho, Vladimir Chigrinov, and Hoi-Sing Kwok, Azo-dyes as photoalignment materials for polymerizable liquid crystals. *Japanese Journal of Applied Physics* **46**, 2995 (2007), Institute of Pure and Applied Physics

be seen that a uniform film can be achieved. Figure 5.12 shows the retardation measurement of this film indicating that it is a positive A-plate with the retardation value as predicted [13].

LCP material UCL017 from Dai-Nippon Ink and Chemicals also shows similar behavior [12]. In fact a photoaligned retardation film and a mechanically aligned retardation film have similar birefringence and optical properties. Again, photoalignment offers advantages such as fine spatial patterning. This will be discussed later in the next section.

Another example of photoaligned retardation film is the negative C-plate. As mentioned above, the VAN mode can be exactly compensated by a negative C-plate (Figure 5.9). The idea is again to use the negative birefringence of the retardation film to compensate for light leakage at oblique angles of incidence. For negative birefringence, the obvious choice is a discotic LC for making such negative C-plates. However, discotic materials are difficult to make. An alternative is to make a film with regular positive uniaxial LCs in a multiple twist (cholesteric) structure. For a cholesteric alignment, it is obvious to see that the effective in-plane

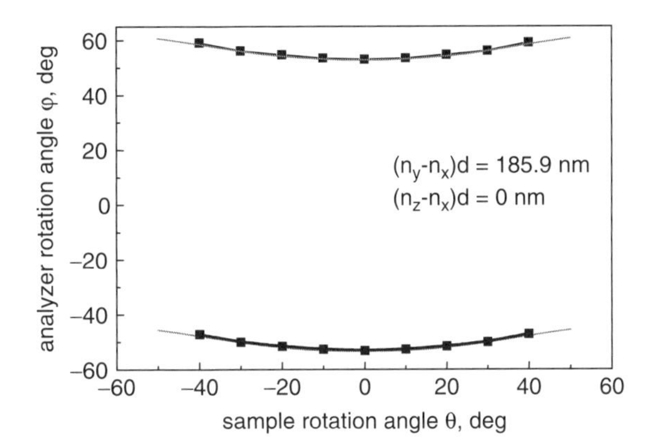

Figure 5.12 Analyzer angle. φ versus sample rotation angle. θ curves (experimental points) measured by transmission null ellipsometry for the optical film shown in Figure 5.11 [13]. The film is optically equal to the positive A crystal plate with a slow axis perpendicular to the polarization direction of the actinic light used in photoalignment processing. The in-plane retardation is 185.9 nm

birefringence is given by $n_x = n_y = (n_e + n_o)/2$ if the number of twists is large, and $n_z = n_o$. Thus, this cholesteric LCP will behave as a negative C-plate, even though a positive nematic LC is used. In fact, a larger positive birefringence for the nematic makes a larger birefringence negative C-plate. Such is the irony of retardation films. Since positive nematic LCP is easily available, this is a better route to the production of negative C-plates.

Yet another method to produce the negative C-plate is to use an amorphous structure. This is strictly speaking not related to photoalignment. But it is an interesting alternative to note anyway. As seen in the previous paragraph, all that is required to make a negative C-plate using uniaxial nematic LC is that the LC molecules should be in-plane. If a homogeneous alignment layer is provided with no rubbing or photoalignment to give it a preferred alignment direction, the LC will be randomly oriented in a so-called amorphous alignment. Amorphous twisted nematic (TN) alignment has been studied in detail [14, 15]. For the negative C-plate, the LC alignment can be TN or homogeneous. Experimentally this type of negative C-plate has not yet been achieved in an LCP. Most researchers still use the cholesteric alignment.

If LCP is used to make a retardation film, then one is not restricted to retardation films with a uniform optical axis. It is possible in fact to have twisted alignment, bend alignment, hybrid alignment, etc., in the same manner as ordinary LCs. After all, these reactive mesogens are LCs before they are polymerized. Thus there is a whole new way to compensate various types of LC modes. In fact, perfect compensation can be achieved when the director alignment of the compensation film is a mirror image of the LC cell itself, and also with a different sign for the birefringence. We shall examine some examples here.

There are several types of non-uniform retardation films in use. An important example is the Fujifilm WV film, which has a hybrid alignment. This film is made from LCP with a discotic structure, which has a negative birefringence. It is excellent for the compensation of TN and optically compensated bend (OCB) modes [4]. In any LCD mode, the most important compensation is for the dark state, as it will affect the contrast ratio directly. For the TN mode, the dark state corresponds to a high-voltage stage with near-homeotropic alignment. But because of strong anchoring, a residual birefringence exists as shown in Figure 5.13.

The wide-viewing-angle (WV) film will exactly compensate this near-homeotropic state for all viewing angles. One can consider the LC near-homeotropic state as two hybrid aligned states with a relative orientation of 90°. For each of these hybrid aligned states, the WV film, which also has a hybrid alignment, will compensate for the residual birefringence of the LC cell since it

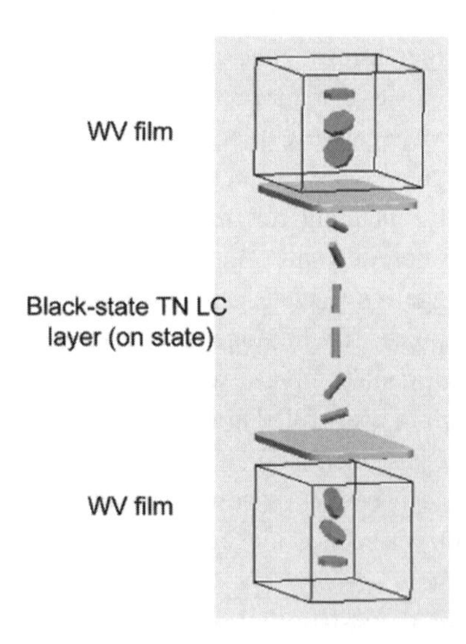

Figure 5.13 Compensation of the dark state in TN mode by discotic negative compensation film

has a negative birefringence. The compensation of the OCB mode follows in an exact manner.

The production of retardation films using LCP generally uses the procedure as indicated in Figure 5.14 [12]. A cell is usually made first with alignment layers on it in the same way as making an LCD cell. The alignment layer of the top substrate has to be weakly bound to the LCP so that it can be removed easily. One is then left with a retardation film on a single substrate. The photoaligned retardation film shown in Figure 5.11 follows this procedure as a matter of fact.

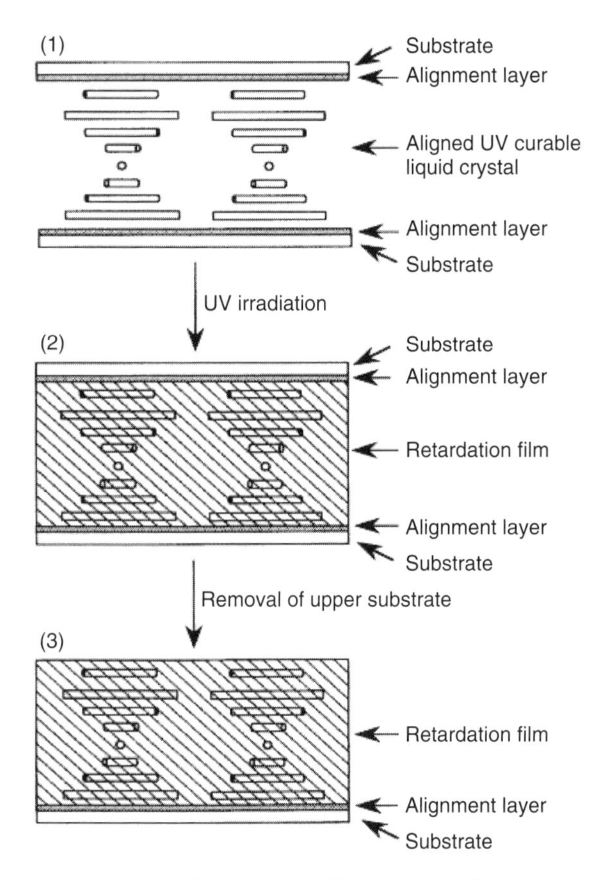

Figure 5.14 The production of retardation films using LCP [12]. Reproduced from H. Hasebe, K. Takeuchi, and H. Takatsu, Properties of UV curable liquid crystals and their retardation. *Journal of the SID* **3**, 139 (1995), The Society for Information Display

This procedure has a drawback in that the removal of the top substrate sometimes will damage the film. It is also not amenable to mass production.

In a more attractive procedure for the production of retardation films using LCP, the top substrate is not used. Instead the LCP mixture is first coated on a single substrate with a suitable alignment layer, either by mechanical rubbing or by photoalignment. The LCP mixture contains a surfactant that will control the alignment angle of the LC molecules at the top free surface. Normally, without any surfactant, the LC molecules will tend to be perpendicular to the free surface. A suitable surfactant will change the LC molecule to be parallel to the free surface. Thus A-plates and C-plates can be made with only one substrate and a free surface. For the preparation of twisted LC structures, a surfactant to provide homogeneous alignment, and appropriate chiral dopants, should be used.

5.2.4 Photo-patterned Phase Retarders and Color Filters

The effect of photo-patterning in polyvinyl-cinnamate (PVCN) derivatives (Figure 2.13) is consequent to the phenomenon of photoinduced optical anisotropy, found during the illumination of PVCN film by polarized UV light of wavelength within its self-absorption band [16]. The main properties of PVCN materials are: (i) a comparatively high value of the photoinduced birefringence (PB) (more than 0.02 at a wavelength of 632.8 nm); (ii) partial reproducibility of the 'writing–erasing' cycles; (iii) long-term memory of the recorded PB value (more than 25 years); and (iv) high spatial resolution ($>8000\ mm^{-1}$). First applications of the effect of photoinduced optical anisotropy in PVCN films comprise [17, 18]: writing polarization holograms and images, planar waveguide optics, polarization optical elements, optical data processing, non-destructive testing of defects, protection of valuable papers as well as other devices by marking them with hidden labels, etc. Most of the early papers devoted to the above-mentioned applications were published in 1977–1987 in the USSR only and are not well known worldwide. The photo-patterned polarization optical elements, obtained by the exposure of PVCN films, are shown in Figures 5.15 and 5.16 [17, 18].

A new impetus in the application of photosensitive media appeared when PVCN was discovered as an LC photoaligning substance [19]. The concept for the design of a photo-patternable hybrid linear photo-polymerizable (HLPP) configuration was proposed by Schadt *et al.* (Figure 5.17) [20, 21]. The HLPP layers include a linearly photo-polymerized polymer (LPP) as the aligning layer and an LPP–LCP phase retarder in one compact implementation (Figure 5.17).

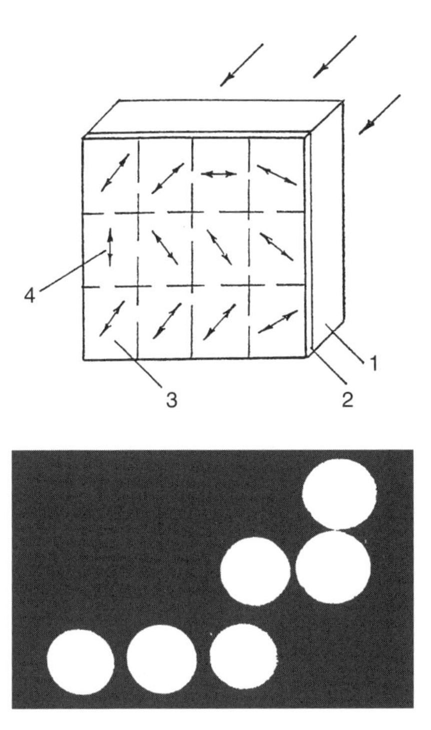

Figure 5.15 Multifunctional polarizer (upper), comprising a linear polarizer (1) and a photo-patterned phase retarder (2), and its image (lower) in crossed polarizers [17, 18]. UV-irradiated (He–Cd laser, $\lambda = 325$ nm) PVCN film (2), used for the photo-patterned phase retarder has a thickness of 10–40 μm. The variation of the optical axis of the phase retarder (3, 4) was done by changing the direction of the polarization of the activated UV illumination. Reproduced from [17, 18], Institute of Physics Publishing, Bristol and Philadelphia, 201–244 (2003)

Figure 5.18 shows one of the configurations of TN-LCD, which makes it possible to obtain birefringent colours, using a photo-patterned phase retarder [18].

The color is made using two phase retarders PR1 and PR2, placed at symmetrical angles α with respect to the input polarizer. TN-LCD (or any other LCD configuration) rotates the 'horizontal' polarization after the P_1 polarizer up to an angle of 90° and then the polarization comes through the 'vertical' polarizer P_2. It can be easily shown that, if light passes through the double phase retarder (DPR) construction with the two symmetrically placed phase retarders PR1 and PR2 with respect to the P_1 followed by the output 'horizontal' polarizer P_3, then the output

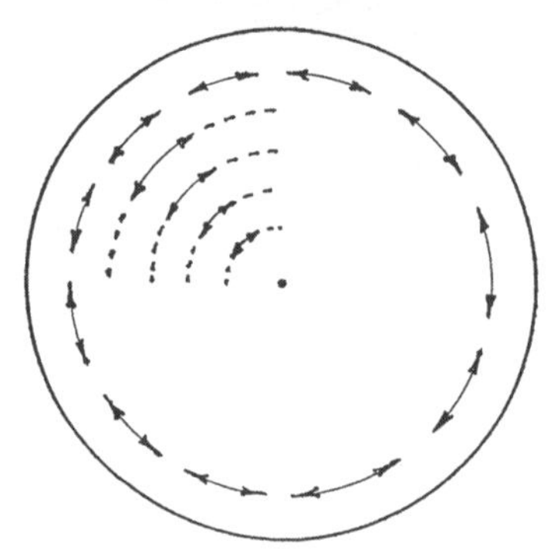

Figure 5.16 Photo-patterned phase plate (upper) and its image (lower) in crossed polarizers [17, 18]. The PVCN plate was rotated at 2 rpm. The activated polarized UV light is along the axis of the plate rotation. The exposure time was 5 min at a power of 10 mW/cm^2. The optical axis was induced in a circular direction and the value of the PB decreases linearly in a radial direction. Reproduced from [17, 18], Institute of Physics Publishing, Bristol and Philadelphia, 201–244 (2003)

intensity I can be written as

$$I = \frac{3}{8} I_0 \sin^2 4\alpha \sin^2 \pi \Delta nd/\lambda \tag{5.8}$$

where I_0 is the light intensity after the input polarizer P_1, Δnd the phase retardation, introduced by each of the two symmetrically placed retarders in the DPR

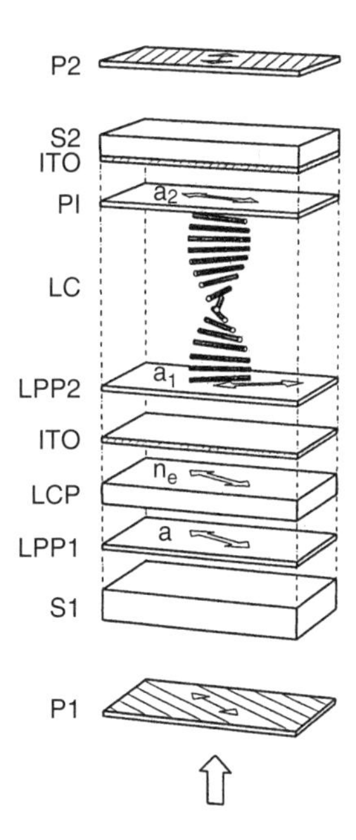

Figure 5.17 HLPP configuration, used for STN-LCD (bottom substrate) [20, 21]. P1, P2, polarizers; LPP1, LPP2, photoaligning layers; LCP, phase retarder; S1, S2, glass substrates; PI, rubbed polyimide layer; ITO, transparent electrodes; a, a_1, a_2, preferred directions of alignment; n_e, optical axis of the phase retarder. Reproduced from [20, 21], Institute of Pure and Applied Physics (1995)

configuration (Figure 5.18), and λ the light wavelength. The angle α should be equal to 22.5° to provide maximum transmission of the light. The appropriate phase retardations Δnd for red, green, and blue light are obtained by setting $\sin^2 \pi \Delta nd/\lambda = 1$, i.e. $(\Delta nd)_{red} = 315$ nm ($\lambda_{red} = 630$ nm), $(\Delta nd)_{green} = 275$ nm ($\lambda_{green} = 550$ nm), and $(\Delta nd)_{blue} = 240$ nm ($\lambda_{blue} = 480$ nm). The advantage of the birefringent color TN-LCD is a low power consumption due to the high transmission in the 'off'-state. The 'on'-state produced by almost homeotropic LCD between crossed polarizers P_1 and P_2 is perfectly dark. The disadvantage of this structure is the angular dependence of the birefringent colors.

To produce wide viewing angles the retarder structure should be more complex, as suggested by Schadt *et al.* [20–22]. Thus two orthogonal positive uniaxial plates behave as a negative phase plate. The plates can be made on the basis of LPP–LCP technology, proposed by Rolic Company (Figure 5.17) [23].

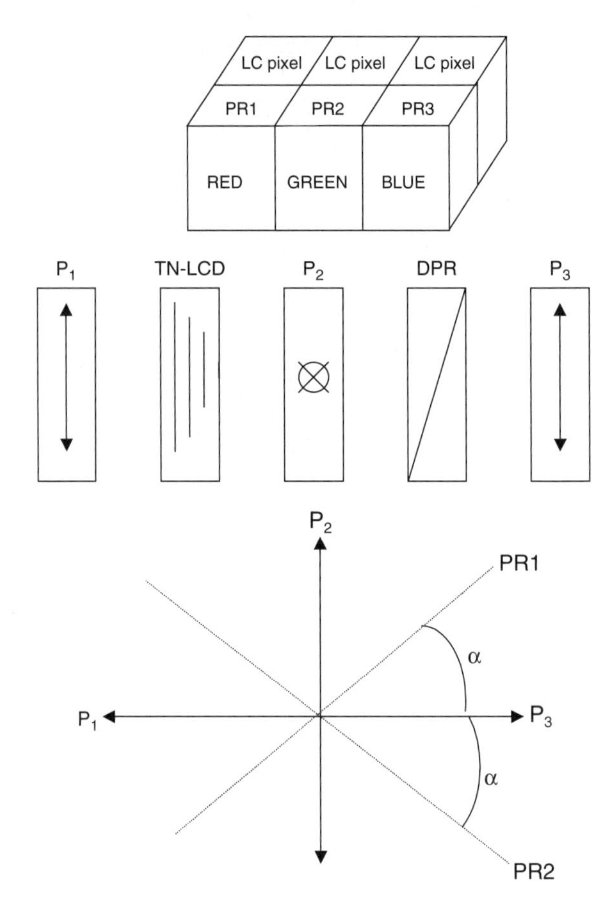

Figure 5.18 New possibility to obtain a color TN- or STN-LCD, using a photo-patterned retarder [18]. Upper: basic red, green, and blue colors obtained by the corresponding properly adjusted phase retarders (PR1, PR2, and PR3). Lower: one of the possible configurations with TN-LCD, three polarizers $(P_1||P_3)\perp P_3$, and double phase retarder (DPR) configuration comprising two phase retarders PR1 and PR2, placed at symmetrical angles α with respect to the input (output) polarizer $P_1||P_3$. Reproduced from [18], Institute of Physics Publishing, Bristol and Philadelphia, 201–244 (2003)

Photo-patterned non-absorbing and bright interference color filters, as well as positive and negative phase retarders based on photoaligning technology, are presented in [22]. The stack of three TN-LCDs, accompanied by yellow, magenta and cyan photo-patterned phase retarders respectively in a front and back surface, can create

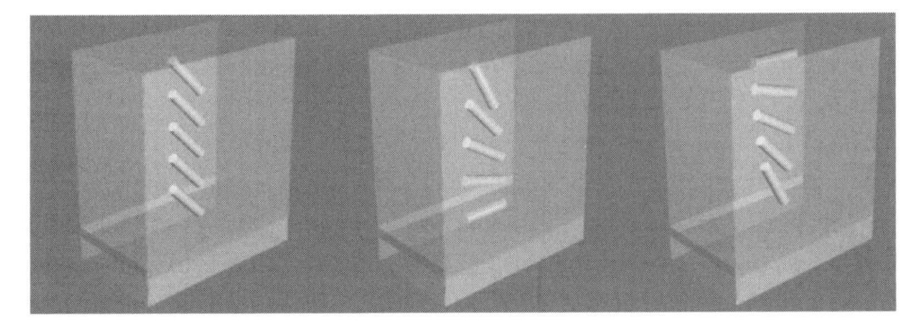

Figure 5.19 Different tilt profiles generated in surface photoaligned LCP retarders. Reproduced from D. K. Yang and S. T. Wu, *Fundamentals of Liquid Crystal Devices*, John Wiley & Sons, Ltd, Chichester, 2006

a full-color LCD, with a transmission that is about six times higher than that of a conventional LCD with absorptive color filters [24]. The manufacture of low-cost, high-brightness birefringent color LCDs, using photo-patterned LCP–LPP phase retarders, can be realized [25]. Rolic's LPP–LCP technology provides a high level of flexibility in the design and manufacture of anisotropic layers. In particular, the process compatibility with flexible substrates and the possibility to adjust the optical axis to any azimuthal and polar angle make it ideal for large-volume roll-to-roll production of retarders and wide-view films for LCDs. The optical performance of such films can be tailored to application-specific requirements by introducing tilt profiles of the optical axis and/or generate patterned retarders with continuous or periodic in-plane variation of the optical axis. Stacking of LCP layers with individual optical functionality leads to compact films with new optical features. Recently, new volume-photoalignable retarders have been developed [26], which combine the aligning and retarder functions in a single material (Figure 5.19).

5.3 Transflective LCD with Photo-patterned Polarizers and Phase Retarders

With the help of simple masks, photoalignment provides a means of patterning polarizers and phase retarders inside an LCD. It offers display designers much more freedom in optimizing the performance of the display. In addition to patterned polarizers and retardation films, photoalignment also allows different areas or subpixels to have different LCD modes. Many types of displays will become possible with the availability of such patterning of the pixels and subpixels. One major possible application of such pattern photoalignment is transflective displays, which work well in the transmissive mode as well as the reflective mode in strong ambient light.

Transflective displays are needed in mobile devices, which have to be functional in both indoor and outdoor environments. Under weak ambient light conditions the display should have the backlight unit turned on and work in the transmissive mode. Under strong ambient light in outdoor situations, the backlight should be turned off and the display should work in the reflective mode. This reflective mode will maintain good contrast regardless of the ambient brightness. In addition such a mode saves power.

The simplest transflective display is of course a simple TN cell with two polarizers. This simple transmissive LCD mode can be made into a transflective display by simply adding a partial reflector between the backlight and the rear polarizer. In the transmissive mode, it is an ordinary TN display; in the reflective mode, the front light passes through the cell twice. However, this design is not optimal as the light beam has to traverse the polarizers four times in the reflective mode, reducing its brightness. In addition, parallax is a serious problem for this kind of reflective mode since the partial reflector is placed outside the LC cell. In-cell reflectors are more desirable.

The most important requirement of a transflective display is that both the reflective and the transmissive modes should have the same grayscales. In fact the transmission voltage curve (TVC) and the reflectance voltage curve (RVC) should be the same, or as close as possible. In other words, the two modes should not compete with each other in brightness, since they should operate at the same time. Another requirement is that both the transmissive and the reflective modes should have good light modulation efficiency. There are many proposals for such transflective displays [27]. They can broadly be classified into two types: (i) designs that make use of the same pixels to work in both modes; (ii) designs that have split pixels so that the reflective and transmissive subpixels are spatially separated. Both approaches have their own advantages and disadvantages.

It is certainly desirable to have the same pixels working in the transmissive as well as in the reflective mode. Such designs will reduce manufacturing cost and simplify system integration. However, it is difficult to achieve. One such proposal is shown in Figure 5.20.

This proposal makes use of the three states available in an AFLC. However, in this design, the reflective mode and the transmissive mode do not actually work

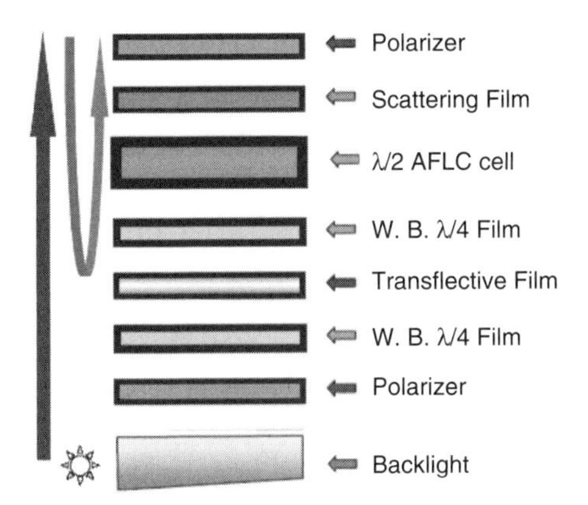

Figure 5.20 The configuration of the transflective display without subpixel separation, using an antiferroelectric LC cell (AFLC) [27]. Reproduced from [27], (2001) The Society of Information Display

with the same LC states. That is, the LC has to be switched to different states for the reflective operation and for the transmissive operation. This is not desirable.

If the subpixels are spatially separated, there are a lot more possibilities and freedom to design the transflective LCD. However, one major requirement for any good transflective display design is that of manufacturability. The reflective and transmissive subpixels should have mostly the same processing steps in order to be easily mass produced. One of the considerations is whether the LC cell has a single cell gap or two different cell gaps for the reflective and transmissive regions. A typical double cell gap structure is shown in Figure 5.21. The reflective subpixel has half the cell gap of the transmissive one. Thus in the ECB mode, both subpixels will have the same birefringence.

From a manufacturing point of view, single cell gap designs are desirable. But in this case, the optical path for the reflective mode is twice that of the transmissive mode. Thus it is very difficult to design such transflective displays to satisfy the optical efficiency requirement. Also, it is rather difficult to match the TVC and RVC. A possible single cell gap design is shown in Figure 5.22.

Thus for a transflective display, one is always faced with compromises, either in complexity of the design in terms of cell gap and subpixel structures, or in complexity of the driving electronics. The situation becomes easier if in-cell patterned retarders or polarizers can be made. An example of an in-cell retardation film display is shown in Figure 5.23. It can be seen that the effect of different cell

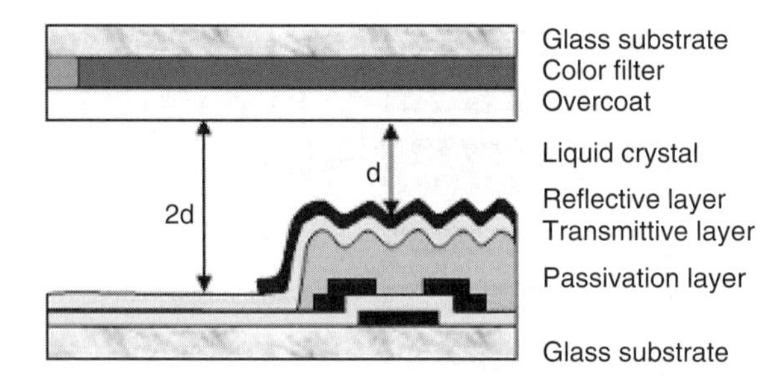

Figure 5.21 A transflective LCD with the double cell gap pixel configuration

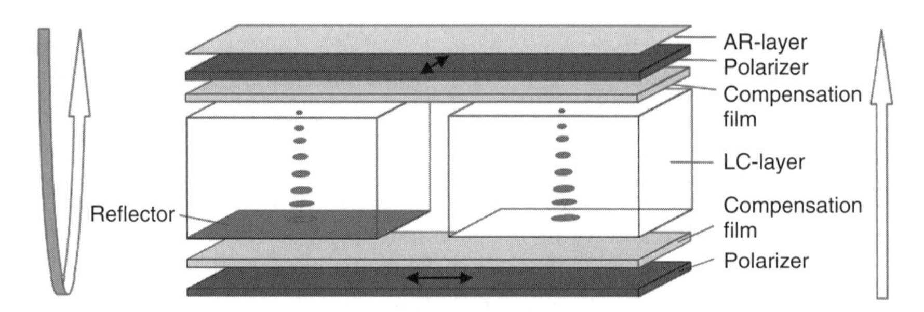

Figure 5.22 The transflective LCD configuration with a single cell gap. Compensation films are optimized for a particular LC electro-optic mode, such as TN or hybrid aligned nematic (HAN) [28]. Reproduced from H. Y. Mak, P. Xu, V. G. Chigrinov, and H.-S. Kwok, Novel single cell gap hybrid aligned nematic transflective LCD. *ASID34, ASID'07 Conference, Singapore* (2007), The Society for Information Display

gaps between the transmissive and the reflective subpixels can be used together with an additional quarter-wave retardation film for the transmissive part. The in-cell quarter-wave retardation films can be made using photoalignment [29–31]. In particular, the basis for the patterned retarder technology can be *in situ* photopolymerization of LC diacrylates, made by Philips [29, 30].

In fact, for the reflective subpixel of a transflective display, a single polarizer is always used. The optics of a single polarizer LCD have been analyzed by many authors [32, 33]. The most common mode is the mixed TN–twist birefringent (MTB) mode. The RVC and the TVC can be matched simply by adding a quarter-wave film, as shown in [32, 33]. Thus, if an internal retardation film can be used, such MTB modes can be used with good optical quality.

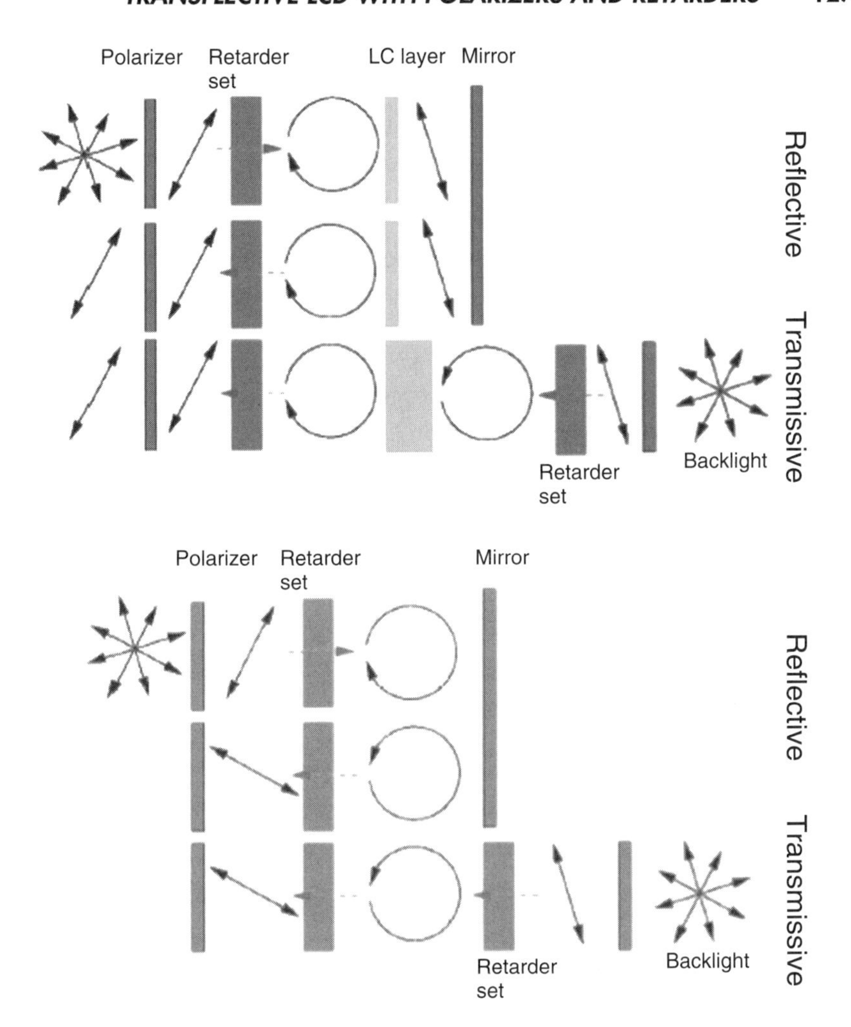

Figure 5.23 The effect of different cell gaps for the transmissive and the reflective subpixels together with an additional quarter-wave retardation film for the transmissive part. The upper picture is the 'off'-state, the lower the 'on'-state (the LC layer keeps the polarization state of the light for sufficiently high voltage). The transmissive part is two times thicker than the reflective part. The LC layer has phase retardation $\lambda/4$ in the reflective part and $\lambda/2$ in the transmissive part

In addition, if it is possible to have patterned alignment layers, different LC modes for the subpixels can be made and optimal performance transflective displays can then be achieved. These novel designs all require patterning of either the polarizer or the retardation film, or the alignment layers. They can all be accomplished with photoalignment [34]. One such novel design that maintains a uniform cell gap is shown in Figure 5.24.

Here the alignment layer is patterned so that in one subpixel, the LC is in the bend alignment, and it is a TN cell in the other subpixel. By adjusting the retardation of the bend cell, it can be seen that this display can function as a reflective display as well as a transmissive display at the same voltages.

Optical retarders are indispensable for making high-performance LCD panels. Recently, a lot of effort has been made to put retarders inside transflective LCDs with the aim of achieving high image quality and reducing the thickness of the panel [35]. In-cell retarders can be fabricated by *in situ* photo-polymerization of LC monomers [36] (Figure 5.25). For this, purpose an alignment layer is essential to align the monomers. A photoalignment layer is thought to be promising because of the high potential for patterning of the optical axis and cutting down on dust compared with the conventional rubbing type of alignment layer. An optimized photoalignment layer was found to be suitable for making in-cell retarders and achieved high thermal stability of the layer comparable with that of a conventional rubbed polyimide alignment layer [36]. Patterning resolution using an optimized photoalignment layer was about 5 μm, while the sensitivity of the photoalignment layer was better than 50 mJ/cm^2, suitable for roll-to-roll processing.

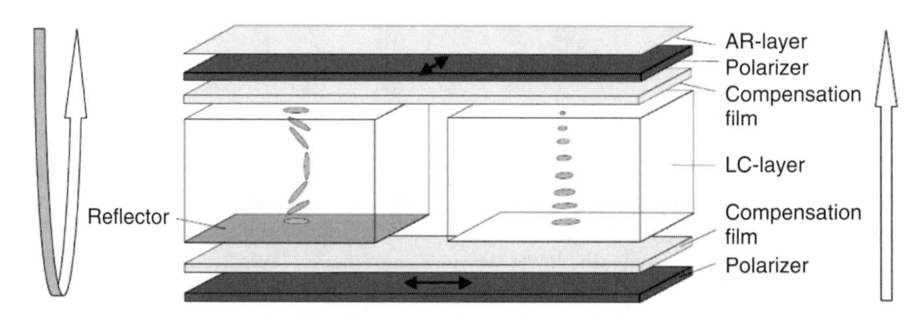

Figure 5.24 The transflective display with a uniform cell gap and two different LC electro-optical modes for reflective (OCB) and transmissive (TN) parts [34]. Reproduced from P. Xu, H. Y. Mak, V. G. Chigrinov, and H.-S. Kwok, Transflective LCD with single cell gap containing two modes in one pixel. *ECLC 2007, Lisbon, Portugal*, Abstracts, PC20 (2007)

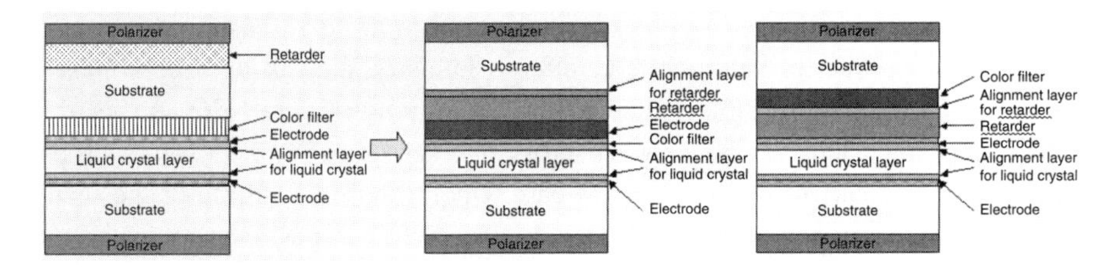

Figure 5.25 Left: structure of conventional LCD with the retarder on top of the LC cell; right, in-cell retarder in different positions with respect to color filters

5.4 Security Applications of Photoaligning and Photo-patterning

A new generation of optical security elements based on photo-patterning technology has enabled the fabrication of optical security devices. The photo-patterned security mark can be applied directly to documents via printing or coating methods as well as indirectly by hot-stamping processes (Figure 5.26) [17, 23].

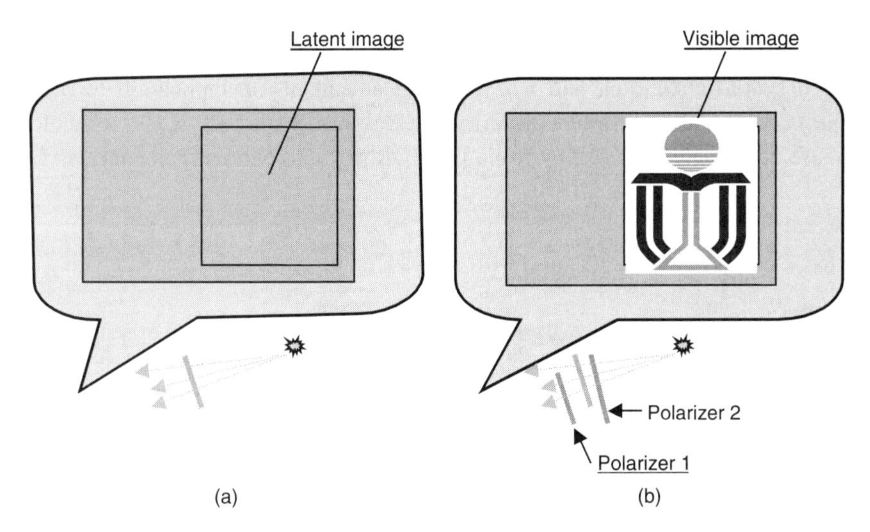

Figure 5.26 Latent (a) and visible color (b) image, produced in a photo-patterning photosensitive layer, such as PVCN or azo-dye material [18]. Reproduced from [18], Institute of Physics Publishing, Bristol and Philadelphia, 201–244 (2003)

A latent image is created on the transparent substrate (e.g. plastic film, paper, glass, clock, product of automobile or aircraft industry, etc.) using a photomask. This latent image is invisible under normal conditions (Figure 5.26(a)), but the image appears when viewed between crossed polarizers (Figure 5.26(b)) or by using a single polarizer in the reflective case. Upon rotation or tilting of the security device, the contrast of the image can be changed from positive to negative or one image can be changed to the other. A color shift and stereo optical effects are possible. The optical security elements cannot be photocopied, but can be machine readable and personalized. The photo-patterning technology of photosensitive layers such as PVCN provides a high resolution security image [18].

Optical security devices using photoaligned alignment-patterned LC cells have been proposed [37]. Latent images with continuous gray levels and high spatial resolution can be written and erased by controlling the azimuthal anchoring strength of the LC alignment film (see Equation (3.2) and Figure 4.8). Two or three latent images can also be created in one guest–host mode LC cell by using one or two polarizers. Moreover, the alignment-patterned LC cell was utilized to visualize the latent information (Figure 5.27).

The latent image with continuous gray levels can be obtained through the process as well as printing on black and white photographic paper. The images can be visualized when using one or two polarizers. The multiple alignment process is very simple and suitable for both large- and small-area patterning with a high spatial resolution, and has the great advantage of application to optical security devices [37]. Similar security devices based on LPP–LCP technology (Figure 5.17) were proposed by Rolic [23] by structured photoalignment of an LPP

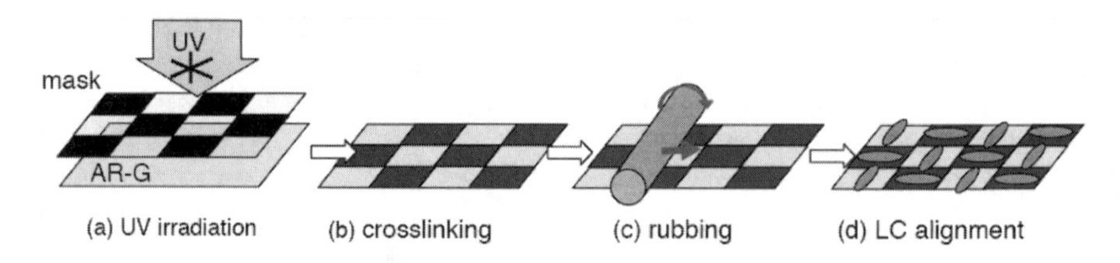

Figure 5.27 The alignment-patterned process on crosslinkable photosensitive polymer AR-G [37]: (a) exposure; (b) crosslinked and uncrosslinked domains; (c) rubbing treatment; (d) LC alignment direction on the AR-G surface. Reproduced from R. Yamaguchi and S. Sato, Liquid crystal optical security device with polarized latent images, *IDW'05 Digest*, p. 65 (2005), Japan

layer through a photomask, thus generating a high-resolution, photo-patterning aligning layer which carries the aligning information of the image to be created.

We have described optically rewritable e-paper in Chapter 4 (Section 4.4). Possible applications of this e-paper concept are light printable and rewritable paper, labels, and plastic card displays, as well as rewritable 3D paper for security applications and overhead projection including building designs, various 3D outline sketches, and maps [38].

5.5 Optical Elements Based on Photoaligning Technology

We have already discussed photo-patterned LC polarizers and phase retarders as well as diffraction grating structures based on photoalignment and photo-patterning. High-resolution diffraction gratings (>1000 lines/mm) are possible with a high surface sensitivity (writing surface energy density as low as 300 mJ/cm^2) [39]. However, some other important optical elements exist, which can also be successfully implemented on the basis of photoaligning technology.

Lin and Fuh [40] established the feasibility of exploiting the surface-assisted photoalignment effect in dye-doped liquid crystal (DDLC) films as spatial filters with controllable polarization in optical signal processing (Figure 5.28). The fabrication relies on the fact that the various intensities of the diffracted orders are responsible for various changes of the polarization state induced by the photoaligned DDLC film. Specific spatial orders in Fourier optical signal processing can be filtered by use of an analyzer placed behind the sample to control the polarization state of the diffracted orders. The image can be modulated for high pass, all pass, and low pass by changing the angle α between the analyzer and the pump beam polarization (Figure 5.28). This filter is easy to fabricate and convenient to use. It therefore has strong potential for practical applications.

As well, Fuh and his colleagues have demonstrated a highly efficient, polarization-independent, and electrically tunable Fresnel lens based on DDLC (Figure 5.29) [41]. The key element is a photomask imbedded with Fresnel zone plate patterns (Figure 5.29).

This photomask has transparent odd zones and opaque even zones made by etching the chromium oxide layer using electron beam lithography. Linearly polarized, diode-pumped solid-state (DPSS) laser light ($\lambda = 532$ nm), which is within the absorption band of the photosensitive methyl-red (MR) dye (Aldrich), was

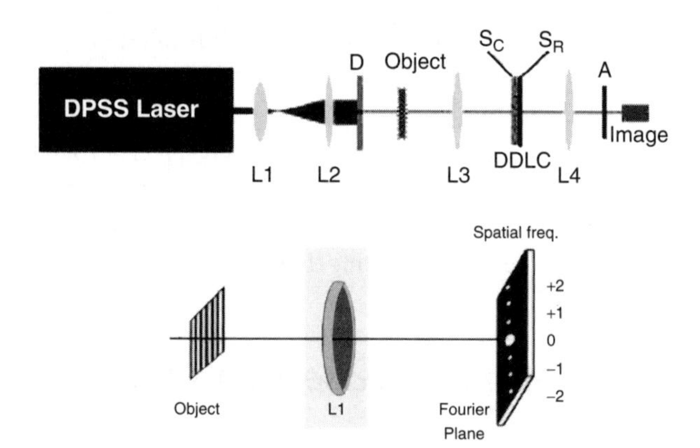

Figure 5.28 Upper: experimental setup [40]. Lenses L1 and L2 are used to expand a plane wave, and D is a diaphragm. Lenses L3 and L4 are the transformation lenses, A is an analyzer, and DPSS a diode-pumped solid-state laser. The DDLC sample is placed in the transform plane. S_C is the command surface that changes the orientation of the photosensitive dye absorbed on it; S_R is the reference surface with rubbed and fixed LC alignment. Lower: the object (a grating mask with 25 mm spacing) is a periodic step function that produces numerous diffracted orders after Fourier transformation; L1, lens. Reproduced from T.-H. Lin and A. Fuh, Polarization controllable spatial filter based on azo-dye-doped liquid–crystal film. *Optics Letters* **30**, 1390 (2005), Optical Society of America

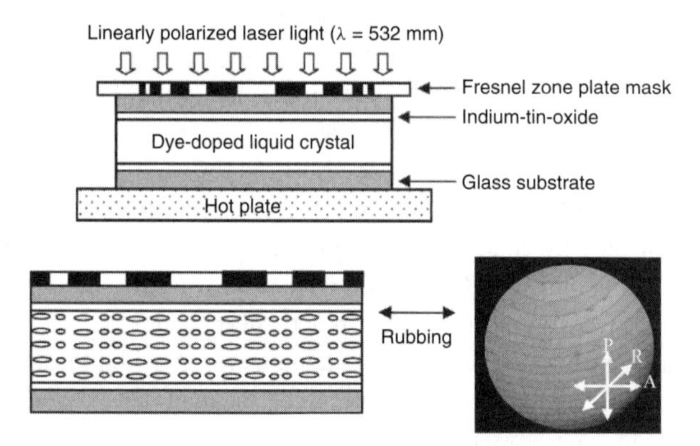

Figure 5.29 Schematic fabrication of the Fresnel zone plate [41]. Also shown is the microscopic image of the fabricated DDLC Fresnel lens observed under a crossed polarizer optical microscope with the rubbing direction (R) of the cell making an angle of 45° with the polarizer axis. Reproduced from L.-C. Lin, H.-C. Jau, T.-H. Lin, and A. Fuh, Highly efficient and polarization-independent Fresnel lens based on dye-doped liquid crystal. *Optics Express* **15**, 2900 (2007), Optical Society of America

used to illuminate the DDLC cell through the Fresnel zone plate mask as shown in Figure 5.29. The polarization of the pump beam was parallel to the rubbing direction of the ITO surface, so the dye molecules tend to align perpendicular to the incident light and the LC molecules in the illuminated parts of the cell surface have turned accordingly, while the previous rubbing direction was kept in black (non-illuminated) zones. As a result, a DDLC Fresnel lens photoaligned with an orthogonally alternating homogeneous binary structure was achieved as shown in Figure 5.29. This orthogonal binary configuration gives the fabricated DDLC Fresnel lens a key property of being independent of the polarization of the incident light. The maximum diffraction efficiency reaches 37%, which approaches the theoretical limit of ~41%. Such a lens functions as a half-wave plate, under the applied voltage. The fabrication of this DDLC Fresnel lens is simple, and the device has fast switching responses (8–18 ms) between the focusing (primary focal distance was 39.5 cm) and defocusing state. Thus, it has a high potential for use in various optical systems.

References

[1] N. Kawatsuki, H. Takatsuka, and T. Yamamoto, Coplanar alignment of mesogenic moiety in photocrosslinked liquid crystalline polymer film containing cinnamoyl groups. *Applied Physics Letters* **75**, 1386 (1999).

[2] O. Yaroshchuk and R. Kravchuk, Bifunctional LCD film. *Molecular Crystals and Liquid Crystals* **454**, 355 (2006).

[3] O. Yaroshchuk, J. Ho, V. Chigrinov, and H.-S. Kwok, Azo-dyes as photoalignment agents for polymerizable liquid crystals. *IDW'06 Digest*, p. 83 (2006).

[4] H. Mori, The Wide View Film for the enhancement of viewing angle of LCDs. *IEEE Journal of Display Technology* **1**, 179 (2005).

[5] E. Peeters, J. Lub, W. P. M. Nijssen, J. Steenbakkers, and D. J. Broer, Thin film polarizers by in-situ photo-polymerization of highly ordered guest-host systems. *Euro Display 2005 Digest*, p. 165 (2005).

[6] Y. Ukai, T. Ohyama, L. Fennell, Y. Kato, M. Paukshto, P. Smith, O. Yamashita, and S. Nakanishi, Current status and future prospect of in-cell-polarizer technology. *Journal of the SID* **13**/1, 17 (2005).

[7] W. C. Yip, H.-S. Kwok, V. M. Kozenkov, and V. G. Chigrinov, Photo-patterned e-wave polarizer. *Displays* **22**, 27 (2001).

[8] O. Yaroshchuk *et al.*, Recent experimental results obtained in HKUST. *www.cdr.ust.hk*, May (2007).

[9] P. Yeh and C. Gu, *Optics of Liquid Crystal Displays*, John Wiley & Sons, Ltd, Chichester (1999).

[10] H.-S. Kwok, X.-J. Yu, and Y.-L. Ho, Extremely broadband retardation films. *SID'07 Digest*, p. 1390 (2007).

[11] S. Morino, A. Kaiho, and K. Ichimura, Photogeneration and modification of birefringence in crosslinked films of liquid crystal polymer composites. *Applied Physics Letters* **73**, 1317 (1999).

[12] H. Hasebe, K. Takeuchi, and H. Takatsu, Properties of UV curable liquid crystals and their retardation. *Journal of the SID* **3**, 139 (1995).

[13] O. Yaroshchuk, J. Ho, V. Chigrinov, and H.-S. Kwok, Azo-dyes as photoalignment materials for polymerizable liquid crystals. *Japanese Journal of Applied Physics* **46**, 2995 (2007).

[14] S. D. Cheng and Z. M. Sun, Amorphous twisted nematic liquid crystal displays: the characteristics and the theoretical considerations. *Liquid Crystals* **19**, 321 (1995).

[15] S. Kobayashi, Current LCD research in Kobayashi Lab. *Workshop on Information Display, Seoul, Korea*, 1 (1995).

[16] E. D. Kvasnikov, V. M. Kozenkov, and V. A. Barachevskii, Birefringence in polyvinyl cinnamate films, induced by polarized light. *Dokladi Akademii Nauk SSSR* **237**, 633 (1977) (in Russian).

[17] V. M. Kozenkov, V. G. Chigrinov, and H.-S. Kwok, Photoanisotropic effects in poly(vinyl-cinnamate) derivatives and their applications. *Molecular Crystals and Liquid Crystals* **409**, 251 (2004).

[18] V. G. Chigrinov, V. M. Kozenkov, and H.-S. Kwok, New developments in photoaligning and photo-patterning technologies: physics and applications. In *Optical Applications of Liquid Crystals*, Ed. L. Vicari, Institute of Physics Publishing, Bristol and Philadelphia, 201–244 (2003).

[19] M. Schadt, K. Schmitt, V. Kozenkov, and V. Chigrinov, Surface-induced parallel alignment of liquid crystals by linearly polymerized photopolymers. *Japanese Journal of Applied Physics* **31**, 2155 (1992).

[20] M. Schadt, Liquid crystal displays and novel optical thin films enabled by photo-alignment, *Molecular Crystals and Liquid Crystals* **364**, 151 (2001).

[21] M. Schadt, H. Seiberle, A. Schuster, and S. M. Kelly, Photo-generation of linearly polymerized liquid crystal aligning layers comprising novel, integrated optically patterned retarders and color filters. *Japanese Journal of Applied Physics* **34**, 3240 (1995).

[22] J. Chen, K. C. Chang, J. DelPico, H. Seiberle, and M. Schadt, Wide-viewing angle photoaligned plastic films for TN-LCDs. *SID'99 Digest*, p. 98 (1999).

[23] *www.rolic.com* (2007).

[24] J. DelPico, K.-C. Chang, and G. P. Sharp, Plastic LCDs to roll? *Information Display* **7**/00, 28 (2000).

[25] H. Seiberle, C. Benecke, and T. Bachels, Photoaligned anisotropic optical thin films. *SID'03 Digest*, p. 1162 (2003).

[26] H. Seiberle, T. Bachels, C. Benecke, and M. Ibn-Elhaj, Volume photoaligned retarders. *IDW'06 Digest*, p. 303 (2006).

[27] S.-C. Kim, W. S. Park, D. W. Choi, J. W. Kang, G.-D. Lee, T.-H. Yoon, and J. C. Kim, Optical configuration for a transflective display mode using an antiferroelectric liquid crystal cell. *SID'01 Digest*, p. 826, 2001.

[28] H. Y. Mak, P. Xu, V. G. Chigrinov, and H.-S. Kwok, Novel single cell gap hybrid aligned nematic transflective LCD. *ASID34, ASID'07 Conference, Singapore* (2007).
[29] B. M. I. Van der Zande, C. Doornkamp, S. J. Roosendaal, J. Steenbakkers, A. Opt Hoog, J. T. M. Osenga, J. J. Van Glabbeek, L. Stofmeel, J. Lub, Y. Iefuji, and L. Weegels, Technologies towards optical foils applied in transflective LCDs. *IDW'04 Digest*, p. 67 (2004).
[30] B. M. I. Van der Zande, J. Lub, H. J. Verhoef, W. P. M. Nijssen, and S. A. Lakehal, Patterned retarders prepared by photoisomerization and photopolymerization of liquid crystalline films. *Liquid Crystals* **33**, 723 (2006).
[31] Y. Kuwana, H. Hasebe, O. Yamazaki, K. Takeuchi, H. Takatsu, V. Chigrinov, and H.-S. Kwok, Optimization of photoalignment layer for in-cell retarders. *IDW'07 Program, Sapporo, Japan*, LCTp5–3 (2007).
[32] D. K. Yang and S. T. Wu, *Fundamentals of Liquid Crystal Devices*, John Wiley & Sons. Ltd, Chichester (2006).
[33] H.-S. Kwok, *Current Liquid Crystal Display Technologies* (Abstracts), HKUST (2007).
[34] P. Xu, H. Y. Mak, V. G. Chigrinov, and H.-S. Kwok, Transflective LCD with single cell gap containing two modes in one pixel. *ECLC 2007, Lisbon, Portugal*, Abstracts, PC20 (2007).
[35] K. J. Kim, H. Y. Lee, S. D. Roh, S. H. Park, B. G. Ahn, B. S. Cheun, J. S. Park, and B. C. Ahn, New structure of IPS mode with in-cell retarder for TV application. *SID'06 Digest*, p. 1158 (2006).
[36] Y. Kuwana, H. Hasebe, O. Yamazaki, K. Takeuchi, H. Takatsu, V. Chigrinov, and H.-S. Kwok, Optimization of photoalignment layer for in-cell retarder. *IDW'07 Digest*, p. 1673, (2007).
[37] R. Yamaguchi and S. Sato, Liquid crystal optical security device with polarized latent images. *IDW'05 Digest*, p. 65 (2005).
[38] A. Muravsky, A. Murauski, V. Chigrinov, and H.-S. Kwok, Novel optical rewritable electronic paper. *IDW'07 Digest*, p. 1681 (2007).
[39] S. Slussarenko, O. Francescangeli, F. Simoni, and Y. Reznikov, High resolution polarization gratings in liquid crystals. *Applied Physics Letters* **71**, 3613 (1997).
[40] T.-H. Lin and A. Fuh, Polarization controllable spatial filter based on azo-dye-doped liquid-crystal film. *Optics Letters* **30**, 1390 (2005).
[41] L.-C. Lin, H.-C. Jau, T.-H. Lin, and A. Fuh, Highly efficient and polarization-independent Fresnel lens based on dye-doped liquid crystal. *Optics Express* **15**, 2900 (2007).

6

Novel LCDs Based on Photoalignment

6.1 Bistable Nematic Displays

Novel, optically bistable, supertwisted nematic LCD configurations have been obtained using photoaligned and photo-patterned substrates. The display was based on switching between two $0-2\pi$ twist configurations. Undesired intermediate states which reduce bistability were effectively suppressed by photo-patterned domains around picture elements which exhibit different azimuthal and zenithal aligning directions. The high degree of stability of the new, domain-stabilized bistable configurations enables the formation of nematic displays with inherent long-term optical memory (Figure 6.1).

The degree of stability of supertwisted bistable twisted nematic (SBTN) displays is shown to improve considerably in nematic twist configurations generated by photoaligned and domain-stabilized bias tilt angle patterns on one or both display substrates. Incorporation of uniaxial optical retarders into SBTN-LCD was shown to result in black and white display operation.

Photoalignment of Liquid Crystalline Materials: Physics and Applications
V. Chigrinov, V. Kozenkov and H.-S. Kwok

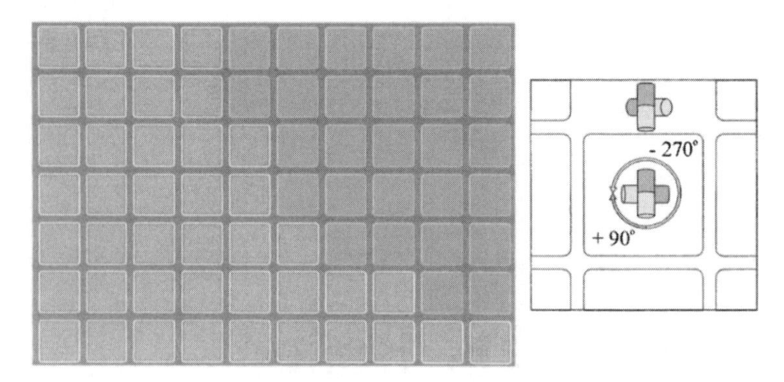

Figure 6.1 Optically bistable, photoaligned, supertwisted nematic LCD [1]. Reproduced from M. Stalder and M. Schadt, Photoaligned bistable twisted nematic liquid crystal displays. *Liquid Crystals* **30**, 285 (2003)

A bistable photoswitchable reflective liquid crystal display (PRLCD) based on azo-dye doped liquid crystal (DDLC), which can be written, erased, and rewritten by changing the polarization of a writing laser beam has been demonstrated [2]. The mechanism is primarily due to the photoinduced adsorption of azo-dyes doped in the LC cell. Linearly polarized diode-pumped solid-state laser light with $\lambda = 532$ nm and an intensity of 150 mW/cm^2 was employed as a writing beam to switch the DDLC sample from a homogeneous texture to a 45° twisted nematic (TN) one (Figure 6.2).

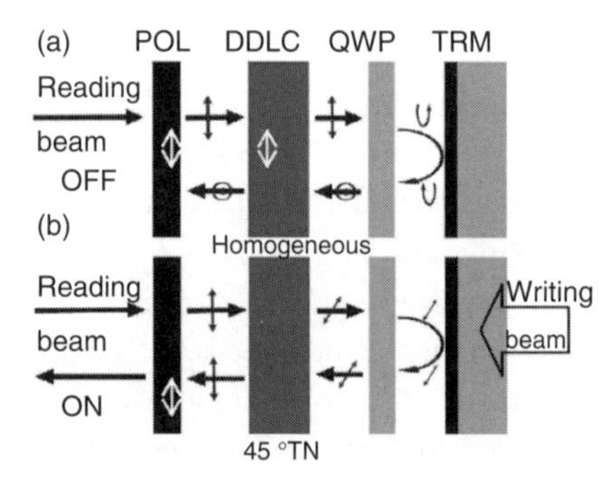

Figure 6.2 A bistable photoswitchable reflective LCD. POL, DDLC, QWP, and TRM are the linear polarizer, dye-doped liquid crystal, quarter-wave plate, and transflective mirror respectively [2]. Reproduced from T.-H. Lin, H.-C. Jau, S.-Y. Hung, H.-R. Fuh, and A. Fuh, Photoaddressable bistable reflective liquid crystal display. *Applied Physics Letters* **89**, 021116 (2006), American Institute of Physics

The polarization plane of the writing beam makes an angle of 45° with the LC director on one of the substrates of the homogeneous cell. The writing beam excites the azo-dyes in the sample. These dyes undergo *trans-cis* isomerization, followed by molecular reorientation and diffusion, and are finally adsorbed onto the surface of the polymer, with their long axes perpendicular to the polarization of the writing beam. The adsorbed dyes then reorient the LC molecules with their directors perpendicular to the directions of polarization and propagation of the writing beam. The LC alignment near the other rubbed polyimide-coated substrate remains unchanged. The anchoring force on the LC molecules exerted by the adsorbed dyes on the polymer surface competes with that exerted by the extension of the alignment effect caused by the rubbed polyimide. Therefore, with sufficient adsorbed dyes, the former overcomes the latter; finally a 45° TN structure is formed. The 45° TN region can be reversibly switched back to homogeneous LC alignment by circularly polarized light. The display provides a reasonable contrast ratio of 10:1 and is easy to fabricate; however, control of the photosensitive dye absorption on the polymer surface still remains a problem.

Truly bistable display cells on plastic substrates with asymmetric surface anchoring conditions on glass and plastic substrates were mentioned earlier (see [46, 47] and Figure 4.12 in Chapter 4). An example of the photoaligned bistable π-BTN LCD is shown in Figure 6.3 [3].

Figure 6.3 Photoaligned π-BTN LCD [3]. Reproduced from F. Yeung and H.-S. Kwok, Truly bistable twisted nematic liquid crystal display using photoalignment technology. *Applied Physics Letters* **83**, 4291 (2003). American Institute of Physics

6.2 Photoaligned Liquid-Crystal-on-Silicon Microdisplays

Liquid-crystal-on-silicon (LCOS) microdisplays are advanced displays that integrate silicon very large-scale integration (VLSI) circuits with thin film transistor liquid crystal displays (TFT-LCDs) [4, 5]. LCOS microdisplays can have very high resolution and yet maintain a large aperture ratio or optical efficiency. As more and more silicon microdisplay panels have been fabricated by submicrometer silicon VLSI technology for higher information content, the pixel pitch has been pushed down to below 10 μm. A compatible LC assembly technology is needed in order to deal with the small dimensions of pixels, particularly for aligning small LC molecules efficiently on such small pixels (Figure 6.4). The conventional process of aligning LC molecules on a silicon panel is by mechanical rubbing. However, the grooves generated by this rubbing are much larger than the LC molecules. The misaligned area may occupy a large portion of small pixels and result in speckles on the display. These small speckles are not observable in a direct-view LCD.

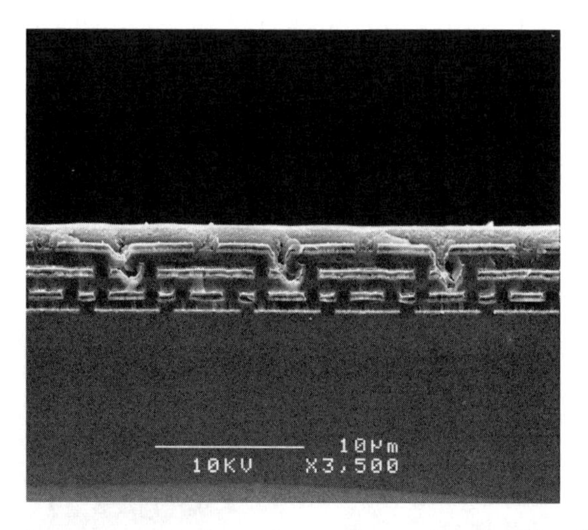

Figure 6.4 Scanning electron microscope image of a cross-section of a silicon panel [4]. Reproduced from B. Zhang, K. Li, V. Chigrinov, H.-S. Kwok, and H. C. Huang, Application of photoalignment technology to liquid-crystal-on-silicon microdisplays. *Japanese Journal of Applied Physics* **44**/6A, 3983–3991 (2005). Institute of Pure and Applied Physics

However, they are not negligible and present a major defect in LCOS microdisplays, where small pixels are projected or magnified for viewing. In addition, mechanical rubbing of a silicon surface may generate dust, electrostatic charges, and scratches and further reduce the yield of LCOS microdisplays. Photoalignment is a non-contact method and shows promise for solving the above-mentioned problems [4, 5]. In addition, high pretilt angles of the alignment layer can be obtained by controlling the illumination angle and dose of polarized UV light. Associated with high pretilt angles are the advantages of the method, namely fast switching time and low threshold voltage.

The photoalignment technology of azo-dye SD1 has been applied to a 0.56 inch LCOS microdisplay panel, which was then assembled on a mixed-mode twisted nematic (MTN) LC configuration for characterization [4]. Both the photoalignment and rubbing methods were used for comparison.

A three-step exposure process was used to improve the quality of LC tilted alignment on the photoaligned surface (Figure 6.5). The two-step exposure, as proposed in Figure 3.4(c), was not efficient, because the reflected non-polarized light tended to destroy the SD1 alignment. The incident angle of the second exposure was optimized to be between 70° and 65°, providing LC pretilt angles of 4.1° and 5.7°, which are comparable with those in rubbed polyimide layers.

The estimated values of the azimuthal and polar anchoring energy of the photoaligned LC layer were 4×10^{-4} J/m^2 and 5×10^{-4} J/m^2 respectively, which are almost comparable with that of the rubbed polyimide layer. Consequently, photoaligned LCOS panels have the same electro-optical characteristics as the usual ones prepared by rubbing (Figure 6.6).

A high contrast and fast response were demonstrated, which meet microdisplay standards. It was also observed that defects were greatly reduced in photoaligned LCOS microdisplays due to the non-contact nature of the photoaligning technology [4].

The high-voltage holding ratio (VHR) of photoaligned microdisplays, greater than 98% at 25 °C, indicated that the azo-dye layer was free from ions causing contamination of the LC mixture (see also Section 3.4). In addition to VHR, residual DC (RDC) voltage charge is another important parameter for silicon microdisplays, which have different electrodes on the silicon the glass substrates. The alignment layer has direct contact with LC mixtures, and silicon pixels might further trap these charges on their surface and increase the RDC charge. RDC voltages for LCOS microdisplays at room temperature were below 50 mV, which was

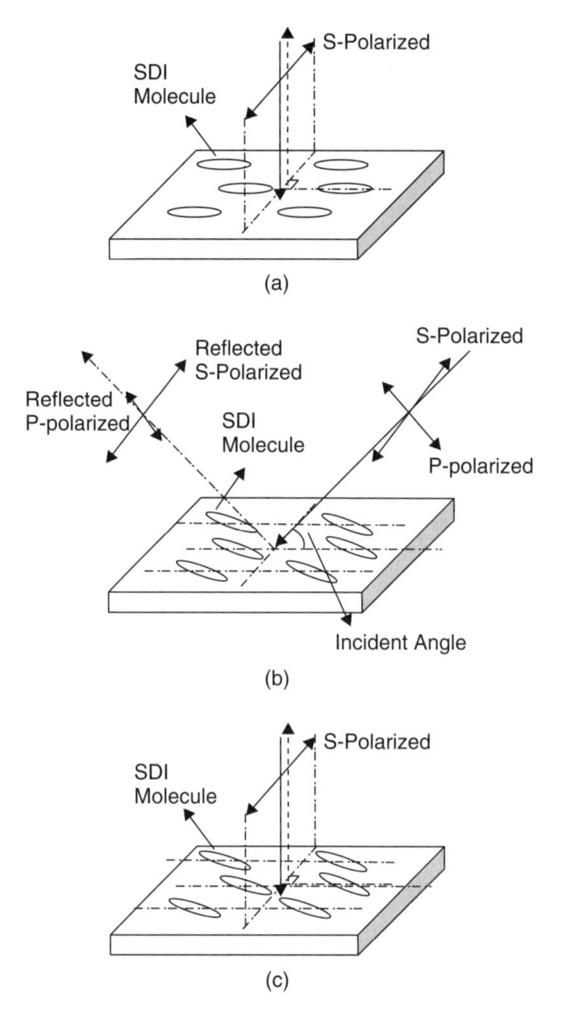

Figure 6.5 Three-step exposure, used for photoaligned LCOS microdisplays [4]: (a) normal exposure; (b) oblique exposure; and (c) normal exposure. Reproduced from B. Zhang, K. Li, V. Chigrinov, H.-S. Kwok, and H. C. Huang, Application of photoalignment technology to liquid-crystal-on-silicon microdisplays. *Japanese Journal of Applied Physics* **44**/6A, 3983–3991 (2005). Institute of Pure and Applied Physics

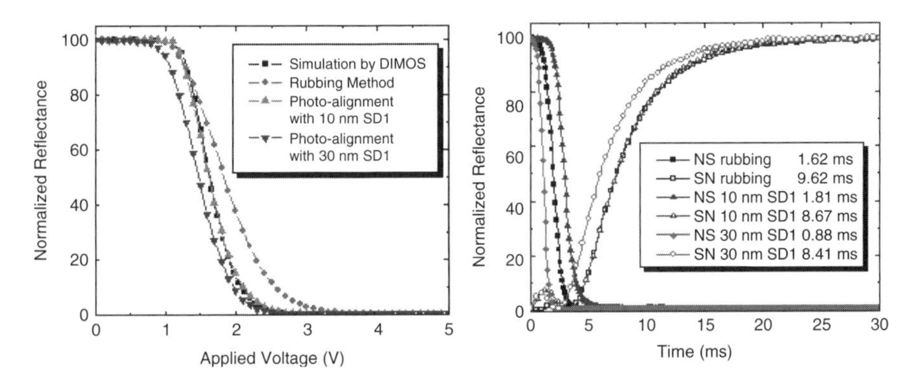

Figure 6.6 Photoaligned LCOS microdisplay electro-optical performance [4]. Comparison of the normalized reflectance (left) and response times (right) between LCOS microdisplays made by rubbing and photoalignment; the results of the simulation using the DIMOS program are also shown. Reproduced from B. Zhang, K. Li, V. Chigrinov, H.-S. Kwok and H. C. Huang, Application of photoalignment technology to liquid-crystal-on-silicon microdisplays. *Japanese Journal of Applied Physics* **44**/6A, 3983–3991 (2005). Institute of Pure and Applied Physics

considered an acceptable RDC voltage for an active matrix LCD (AM-LCD). Thus photoalignment technology may provide a new way for high-quality microdisplays.

6.3 Photoaligned Ferroelectric LCDs

The photoalignment of ferroelectric LCs was analyzed in Section 4.3. We shall describe here a ferroelectric LCD (FLCD) based on photoalignment in more detail. Ferroelectric liquid crystal (FLC) has attracted great interest with regard to passive matrix displays due to its memory effect, wide viewing angles, and fast response time [6]. However, the generation of grayscale has always been an issue [7]. Common surface-stabilized FLC structures with bistable switching can memorize black and white states, but cannot provide an intrinsic grayscale. It is well known that a digital grayscale can be achieved by combining temporal dither (TD) and spatial dither (SD), but it is difficult to achieve more than three gray levels even when SD is used [8]. In order to achieve additional analog gray levels, the multithreshold method (MTM) has been proposed and driving techniques for MTM have been investigated [9]. An intrinsic grayscale can be obtained in the so-called 'thresholdless' FLC structure but the grayscale cannot be memorized [10]. Deformed helix

FLC exhibits a natural grayscale with many gray levels, but the FLC switching in this case shows a 'quasi-bistable' behavior with hysteresis, because the helix formation is hindered [11].

An important characteristic of the FLC effect is charge compensation. On application of an electric field, the LC director in some regions flips into the opposite allowed orientation, and domains of the opposite switched state are formed. Removal of the field during this switching process leaves the device in a multidomain or partially switched state. Hartmann described this charge-controlled phenomenon in FLC samples, which can be observed optically as domains nucleate grow and stabilize, resulting in a controllable, continuous gray shading between black and white [12]. The domains normally form at the same places each time a device is switched, indicating that they are seeded in some way. If we could efficiently control these domains, we would have more gray levels than anyone could use, i.e., we would have a perfect analog rendition of half-tones. Although the working principle seems to be easy, it is limited in depth and too difficult to implement in practice. It shows both lack of reproducibility in time and local variation in space and these methods have not been judged sufficiently practical to compete with other methods in passive driving at present [13].

The memorized grayscale of a passively addressed FLCD can be obtained, if the FLC possesses a high spontaneous polarization $P_s > 50$ nC/cm^2 [14]. However, a depolarizing field appears in FLC cells with high P_s values, suppressing the bistability if the aligning layers are thick enough [7]. Therefore, extremely thin aligning layers are more desirable for the development of a passively addressed FLCD with memorized grayscales.

As shown in Section 4.3, high uniformity of the FLC layer was obtained by a photoalignment technique using the sulfonic azo-dye SD1, whose chemical structure is shown in Figure 2.4. The thickness of the photoaligned layer can be optimized and perfectly suits all the conditions for an FLCD with a memorized grayscale (see Figures 4.5–4.7). Intrinsic grayscale generation and stabilization in FLCD have been proposed and investigated [15]. The dynamic current and electro-optical response in a photoaligned FLC testing cell were developed for grayscale stability against different crosstalk on passive matrix driving. Finally, a passive 160×160 multiplex addressing photoaligned 5 μm reflective FLCD with a high contrast and four memorized grayscale levels was demonstrated, which shows that the grayscales can be accurately reproduced and stabilized [15].

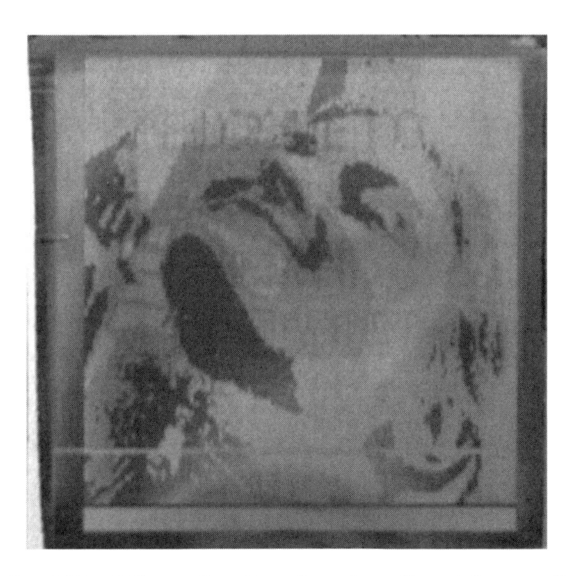

Figure 6.7 Passive 160 × 160 matrix addressing photoaligned 5 μm reflective FLC display (48 mm × 46 mm) with high contrast and four memorized grayscale levels [15]. Reproduced from X. H. Li, A. Murauski, A. Muravsky, P. Z. Xu, H. L. Cheung, and V. Chigrinov, Gray scale generation and stabilization in ferroelectric liquid crystal display. *Journal of Display Technology* **3**, 273 (2007). IEEE

The background of the image in Figure 6.7 looks green in reality, because the optimal conditions of FLCD are not optimal for a black and white (achromatic) appearance. A new configuration of FLCD using two parallel-aligned FLC cells with crossed directors to overcome the color dispersion has been proposed [16]. The idea is to use a compensating FLC cell to cancel the phase retardation caused by the incident light after traversing through the principal cell. Both the principal cell and the compensating cell operate in the surface-stabilized FLC electro-optical mode, and the two layers are nearly identical except for the cell gap. The two cells have same voltage control, thus they can switch at the same time so that the directors of the two cells are always perpendicular. High quality switched FLC cells were obtained by photoalignment technology using azo-dye SD1 (Figure 2.4). The configuration of our experiment for the double cell was as follows. The director of the first cell serving as the principal cell is parallel to the optical axis of the polarizer. The compensating FLC cell and the principal cell are aligned parallel to each other, but the directors of the two FLC cells are crossed. The analyzer

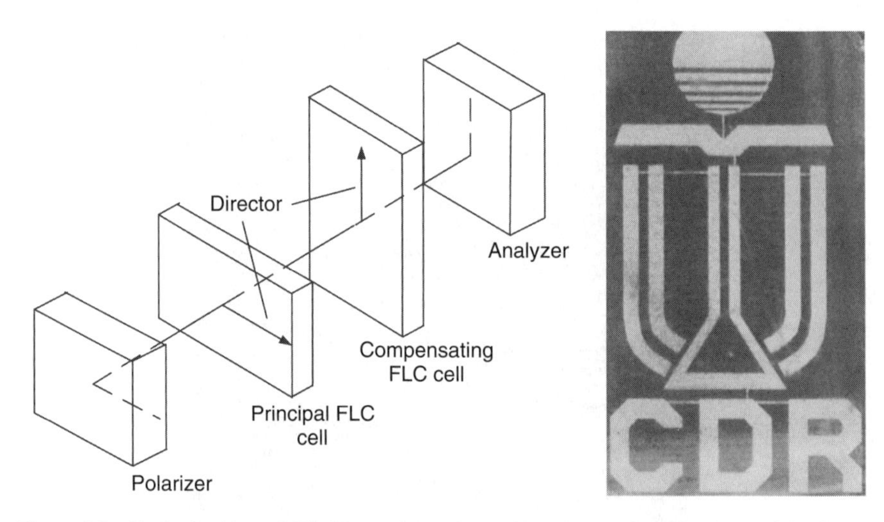

Figure 6.8 Left: double cell FLCD configuration with achromatic switching. Right: image of almost achromatic photoaligned FLCD [16]. Reproduced from P. Xu, X. Li, and V. Chigrinov, Double cells achromatic ferroelectric liquid crystal displays using photoalignment technology. *Japanese Journal of Applied Physics* **45**, 200 (2006). Institute of Pure and Applied Physics

was perpendicular to the polarizer. Figure 6.8 illustrates the configuration for the transmissive mode. The configuration of the reflective mode is similar with the analyzer replaced by a mirror. Almost achromatic switching of a double FLCD cell configuration of any thickness can be obtained by this method (Figure 6.8).

In devices with low power consumption, TFT technology is not widely used due to its relatively high power consumption and high cost. In this context, passively addressed chromatic bistable displays become attractive in many applications, such as e-books, PDAs, and mobile phones with low power consumption. Since monochromatic displays can satisfy most requirements of low-power devices, the use of passive monochromatic FLC bistable display can further reduce the LCD cost and provide better optical performance. The optical parameters of a photoaligned bistable FLCD with monochromatic switching have been optimized using the optical retardation of the LC layer, the cone angle of molecules, and orientations of the polarizers and the retardation films, if any [17, 18]. Several types of inexpensive bistable photoaligned monochromatic FLCs with low power consumption and memorized colors have been demonstrated (Figure 6.9) [18].

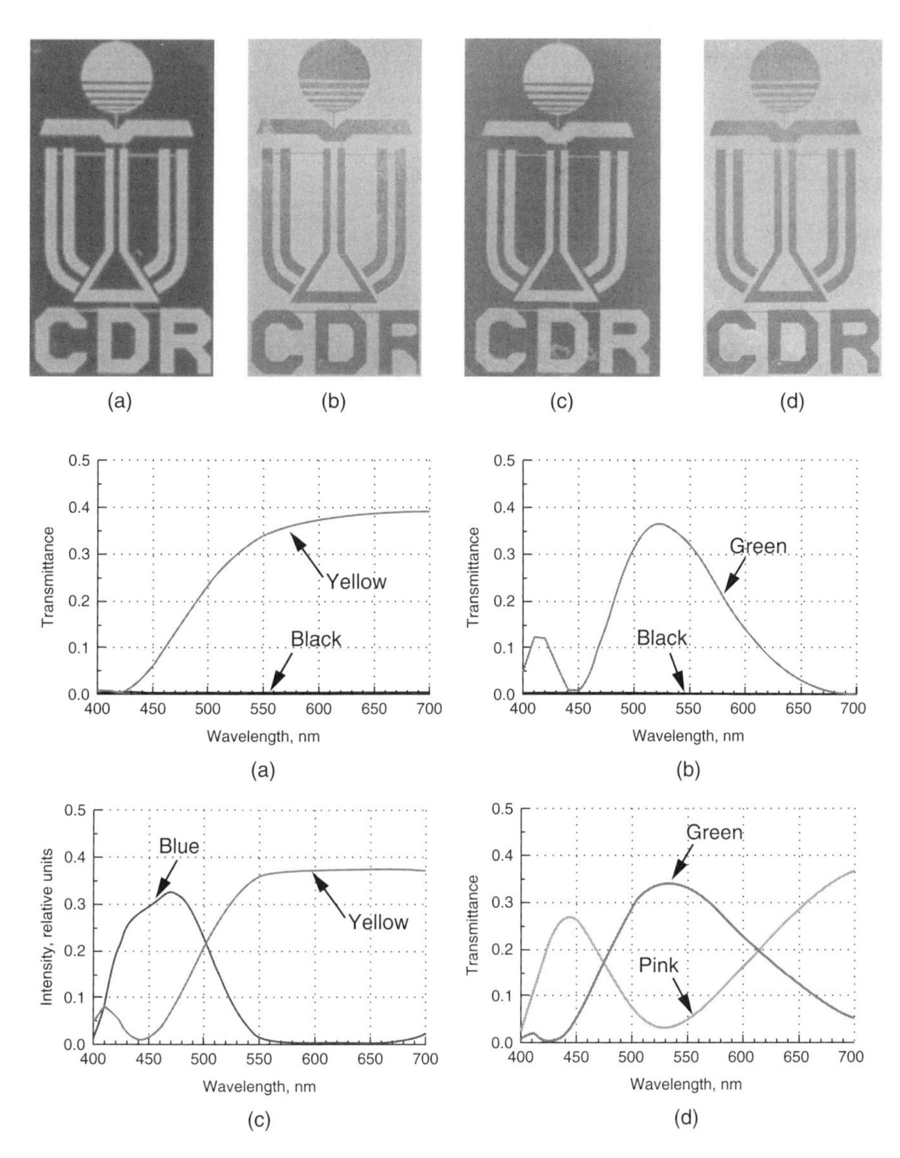

Figure 6.9 Photoaligned bistable FLC displays with memorized colors: upper, the experimental prototypes; lower, spectra of bistable FLC displays [17, 18]. Reproduced from S. Valyukh, I. Valyukh, P. Xu, and V. Chigrinov, Study on birefringent color generation for a reflective ferroelectric liquid crystal display. *Japanese Journal of Applied Physics* **45,** 7819 (2006). Institute of Pure and Applied Physics

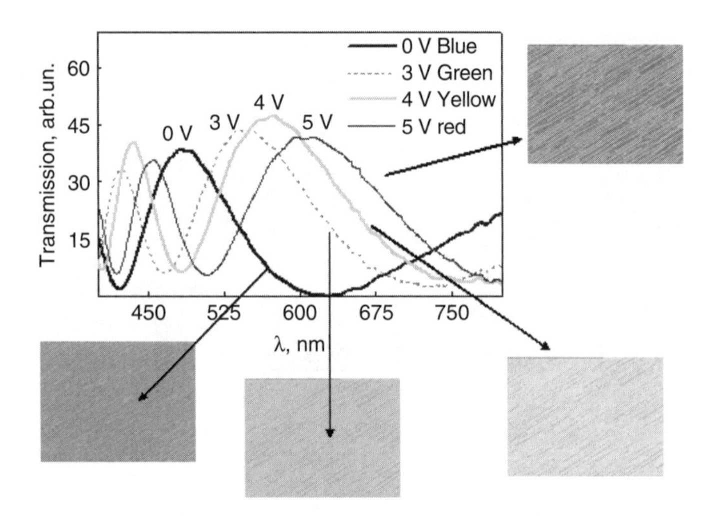

Figure 6.10 Light transmission spectra of 5 μm DHF-FLC cell placed between two crossed polarizers versus applied voltage at a frequency of color switch of 100 Hz; the horizontal size of the micrographs is 500 μm [19]. Reproduced from E. P. Pozhidaev, G. Hegde, P. Xu, and V. G. Chigrinov, Electrically controlled birefringent colors of smectic C* deformed helix ferroelectric liquid crystal cells. *FLC'07 Program, FLC International Conference on Ferroelectric Liquid Crystals*, O-36 (2007)

Similar birefringent color deformed helix ferroelectric (DHF) FLC can be obtained if high-quality photoaligned azo-dye films are used (Figure 6.10). The change of color is done by varying the driving voltage amplitude from 0 to 5 V at a frequency of 5–500 Hz for 5 μm photoaligned DHF-FLC display cells with a helix pitch of less than 1 μm [19]. This principal of efficient color generation looks promising for a field sequential FLCD without color filters.

As we reported in Section 4.3, a high-quality, field sequential, full-color polymer-stabilized FLCD exhibiting V-shape switching (PSV-FLCD) has been successfully implemented by Kobayashi *et al.* using photoalignment technology (see [29, 30] in Chapter 4).

6.4 New Optical Rewritable Electronic Paper

The idea of electronic paper is to store and display information generated by computer on a lightweight thin, flexible, and robust paper-like carrier with good brightness, high contrast, and full viewing angle. Ultra-low or zero power consumption is highly desirable [20]. Several LC-based candidates for electronic paper have been announced. The most mature are the cholesteric LCD [21], zenithal bistable device (ZBD) [22], and bistable nematic (BiNem) [23]. However, electronic paper application of LCs still suffers from high-level complexity of the driving electronics, which often fail for flexible display due to the insufficient durability of the flexible conductor and contact bonding. Besides, the recent jump in the price of indium, which is mainly used for the creation of the transparent conductor layer (ITO), has also increased the cost of all ITO-containing devices.

Recent developments in optically rewritable (ORW) LC photoaligning display and progress in LC photoalignment have made it possible to separate the e-paper display unit and the driving optoelectronics part, together with a significant reduction in complexity of our ORW e-paper structure making both device properties and cost paper-like [24, 25]. That makes this ORW e-paper very durable and cheap and ready for the challenge of flexibility. Further investigations of regimes of operation [24] have allowed us to use cheap and low-power-consuming 450 nm LEDs as an alternative exposure light source, instead of expensive and high-power-consuming mercury lamps or lasers. The device structure of ORW electronic paper in the experimental prototype based on a polarizer and plastic substrates has been successfully implemented [25] (Figure 6.11). It operates by ORW alignment technology, has no electrodes, possesses grayscale capability, is truly stable, and does not require power to show the image which has a wide viewing angle and high contrast. This new electronic ORW paper display is very tolerant to cell gap variation, since even a 50% variation in the cell gap will not cause a noticeable change in the LC transmission value [24, 25].

In addition to this new single side-light printable ORW e-paper for 2D images, a new function for 3D images on e-paper in LC technology has been developed [25] by combining two ORW LC layers in one unit to obtain e-paper with advanced security features or even a stereoscopic 3D image (Figure 6.12).

For the realization of a stereoscopic 3D image, the display should be able to show independent pictures in orthogonally polarized light for the left and the right eyes. The new e-paper consists of two LCDs based on ORW technology, where a 2D image can be seen directly and a 3D image can be seen with passive polarizer glasses (Figure 6.12). The first example of the practical implementation of the 3D structure of new photoaligned e-paper using two 10 μm LC cells filled with MLC-6080 from Merck is shown in Figure 6.13.

Figure 6.11 Left: single side-light printable ORW e-paper (side views at 45°). Right: ORW paper with improved dark and bright states [25]. Reproduced from A. Muravsky, A. Murauski, V. Chigrinov, and H.-S. Kwok, Novel optical rewritable electronic paper. *IDW Conference Digest*, p. 1681 (2007), Sapporo, Japan

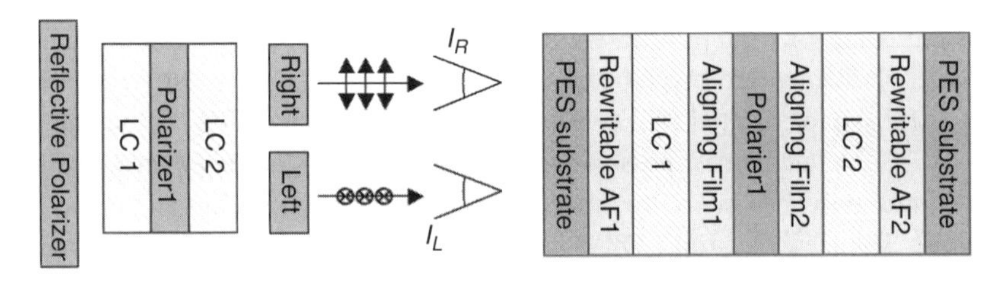

Figure 6.12 Optical scheme (left) and structure (right) of double side printable ORW e-paper for a 3D image [25]. Reproduced from A. Muravsky, A. Murauski, V. Chigrinov, and H.-S. Kwok, Novel optical rewritable electronic paper. *IDW Conference Digest*, p. 1681 (2007), Sapporo, Japan

Possible but not limiting applications of the new e-paper based on photoaligning are lightweight printable rewritable paper, labels, and plastic card displays, as well as rewritable 3D paper for security applications and overhead projection including building designs, various 3D outline sketches, and maps.

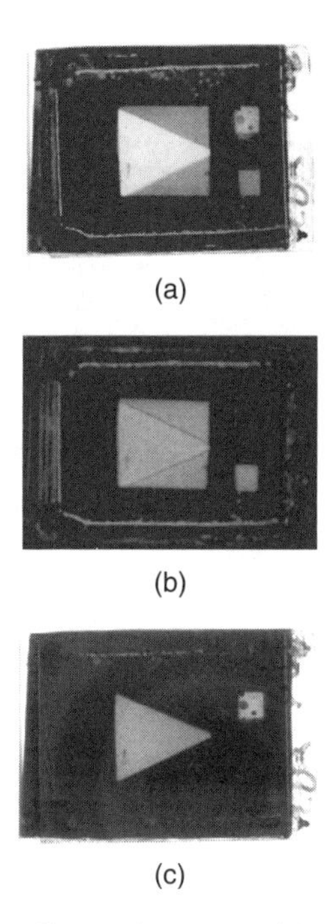

(a)

(b)

(c)

Figure 6.13 ORW e-paper test cell (on glass): (a) without polarizer glasses (2D image); (b) through left polarizer; (c) through right polarizer [25]. Reproduced from A. Muravsky, A. Murauski, V. Chigrinov, and H.-S. Kwok, Novel optical rewritable electronic paper. *IDW Conference Digest*, p. 1681 (2007), Sapporo, Japan

6.5 Application of Photoalignment in Photonic LC Devices

Certain examples of LC photoalignment, useful for photonic LC devices, were mentioned earlier in Sections 4.7 and 4.9. Perfect, uniform, homogeneous LC alignment can be made inside cylindrical photonic holes with a submicrometer diameter (see Figure 4.23) or on top of the silicon microring system (see Figure 4.25). If the flat areas of the silicon chip (see e.g. Figure 4.25) are kept untreated, the absence of LC alignment direction, which can only come from the

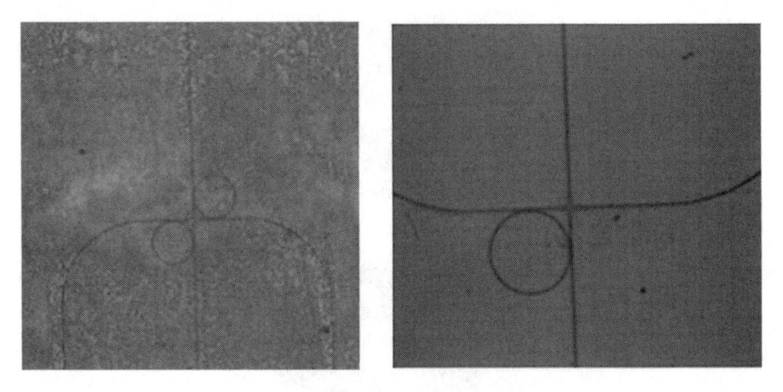

Figure 6.14 Microscopic picture of assembled LC-clad photonic chip in reflected non-polarized light. Left: without alignment layer (with schlieren structure). Right: with azo-dye SD1 photoalignment layer on bottom substrate [26]. Reproduced from A. Muravsky, Optical rewritable photoalignment technology: application in photonics and displays. PhD thesis proposal, HKUST (2008)

surface, gives rise to other kinds of defects, so-called schlieren structure types [26]. Such defects were observed when the profiled silicon substrate was used without any alignment treatment (Figure 6.14).

Poor LC alignment on the uneven surface of a photonic chip is one reason for the considerable reduction in the overall LC refractive index change in the cladding under an external electric field and for consequent inefficient operation of the LC photonic device [27]. This problem can be overcome when a thin 12–14 nm

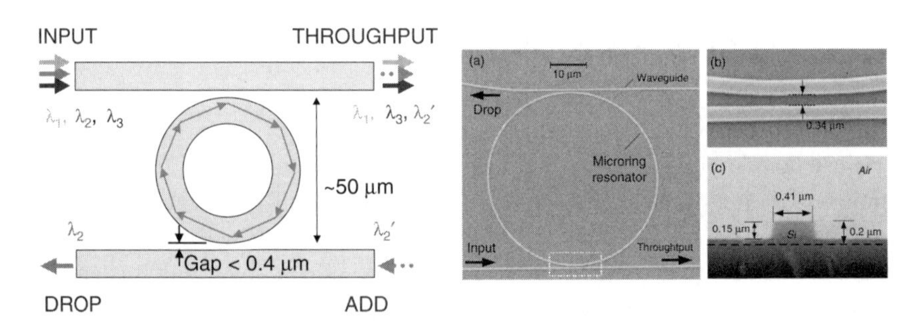

Figure 6.15 Microring resonator based on Si technology [28, 29]. Reproduced from A. Poon, L. Chao, M. Ning, L. S. Lik, D. Tong, and V. Chigrinov, Photonics filters, switches and subsystems for next-generation optical networks. *HKIE Transactions* **11**, 60 (2004), HKIE

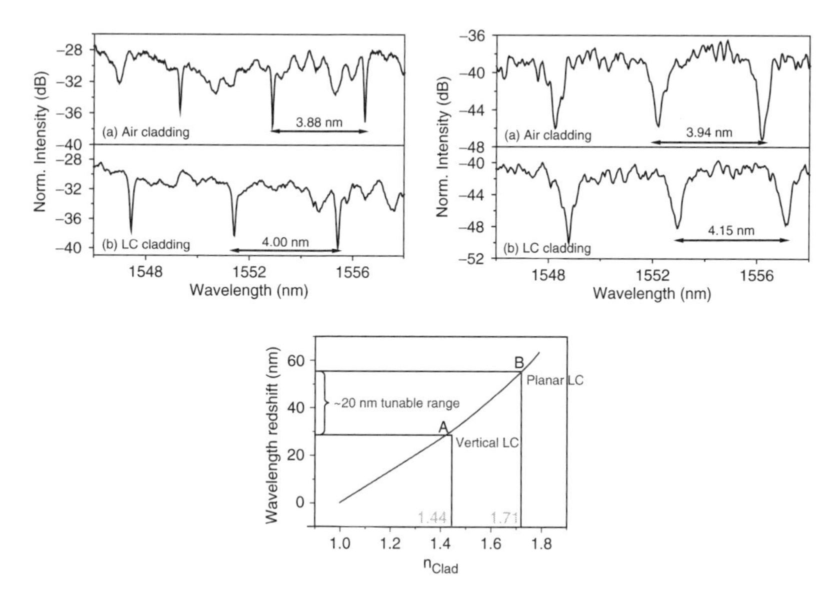

Figure 6.16 Characteristics of an electrically tunable microresonator using photoaligned LC as the cladding layer shown in Figure 4.25. From left to right: FSR shift when air cladding is changed to planar oriented LC; FSR shift when air cladding is changed to homeotropic oriented LC; wavelength redshift as a function of the effective refractive index of the LC layer [29]. Reproduced from V. G. Chigrinov, L. Zhou, A. A. Muravsky, and A. W. Poon, Electrically tunable microresonators using photoaligned liquid crystals as cladding layers. US Provisional Patent Application No. 60/794,128, filed on April 24, 2006, US Patents Office

azo-dye photoalignment layer is spin coated on top of the silicon substrate and the alignment direction selected by photoexposure to polarized light (Figure 6.14). Thus photoalignment technology, which enables us to solve this problem, becomes very important.

An add/drop Si-based photonic filter is shown in Figure 6.15. The operating principle is based on the resonance effect in the microring, which can drop one of the input wavelengths coming through the filter. Switchable filters are highly desirable and can be implemented using photoaligned LC cladding on top of the microring surface (see also Figure 4.25).

LC alignment on Si surfaces with submicrometer-sized straight and curved waveguide profiles has been studied [28, 29]. The LC cladding refractive index was varied by an applied voltage, and subsequently the microresonator resonance wavelengths were tuned. Based on our initial measurements, the free spectral range (FSR) wavelength shift within the range of 20 nm was obtained, which is comparable with a thermo-optic effect (Figure 6.16). New voltage-controllable Si-based add/drop filters are envisaged from this principle.

Thus photoaligning technology has obvious advantages in comparison with the usual 'rubbing' treatment of the substrates for application in LC photonic devices. Possible benefits from using this technique in photonic LC devices include [30]: (i) new advanced applications of LC in fiber communications, optical data processing, holography, and other fields, where the traditional rubbing LC alignment is not possible due to the sophisticated geometry of the LC cell and/or high spatial resolution of the processing system; (ii) the ability for efficient LC alignment on curved and flexible substrates; and (iii) manufacturing of new optical elements for LC technology, such as patterned polarizers and phase retarders, tunable optical filters, variable optical attenuators, etc.

References

[1] M. Stalder and M. Schadt, Photoaligned bistable twisted nematic liquid crystal displays. *Liquid Crystals* **30**, 285 (2003).

[2] T.-H. Lin, H.-C. Jau, S.-Y. Hung, H.-R. Fuh, and A. Fuh, Photoaddressable bistable reflective liquid crystal display. *Applied Physics Letters* **89**, 021116 (2006).

[3] F. Yeung and H.-S. Kwok, Truly bistable twisted nematic liquid crystal display using photoalignment technology. *Applied Physics Letters* **83**, 4291 (2003).

[4] B. Zhang, K. Li, V. Chigrinov, H.-S. Kwok, and H. C. Huang, Application of photoalignment technology to liquid-crystal-on-silicon microdisplays. *Japanese Journal of Applied Physics* **44**/6A, 3983–3991 (2005).

[5] H. Selberle, O. Muller, G. Marck, and M. Schadt, Photoalignment of LCoS LCDs. *Journal of the SID* **10**/1, 31 (2002).

[6] N. A. Clark and S. T. Lagerwall, Submicrosecond bistable electrooptic switching in liquid crystals. *Applied Physics Letters* **36**, 899 (1980).

[7] V. G. Chigrinov, *Liquid Crystal Devices: Physics and Applications*, Artech House, London (1999).

[8] S. T. Lagerwall, *Ferroelectric and Antiferroelectric Liquid Crystals*, John Wiley & Sons, Weinheim (1999).

[9] A. Tagawa, P. Bonnett, and T. Numao, Driving scheme for ferroelectric liquid crystal display with multithreshold method by thickness variation. *Japanese Journal of Applied Physics* **43**, 7150 (2004).

[10] L. M. Blinov, E. P. Pozhidaev, F. V. Podgornov, S. A. Pikin, S. P. Palto, A. Sinha, A. Yasuda, S. Hashimoto, and W. Haase, 'Thresholdless' hysteresis-free switching as

an apparent phenomenon of surface stabilized ferroelectric liquid crystal cells. *Physical Review E* **66**, 021701 (2002).

[11] L. A. Beresnev, V. G. Chigrinov, D. I. Dergachev, E. P. Pozhidaev, J. Funfschilling, and M. Schadt, Deformed helix ferroelectric liquid crystal display: a new electrooptic mode in ferroelectric chiral smectic C liquid crystal. *Liquid Crystals* **5**, 1171 (1989).

[12] W. Hartmann, Charge-controlled phenomena in the surface-stabilized ferroelectric liquid-crystal structure. *Journal of Applied Physics* **66**, 132 (1989).

[13] E. Pozhidaev, V. Chigrinov, and X. Li, Photoaligned ferroelectric liquid crystal passive matrix display with memorized gray scale. *Japanese Journal of Applied Physics* **45**, 875 (2006).

[14] E. P. Pozhidaev and V. G. Chigrinov, Bistable and multistable states in ferroelectric liquid crystals. *Crystallography Reports* **51**, 1030 (2006).

[15] X. H. Li, A. Murauski, A. Muravsky, P. Z. Xu, H. L. Cheung, and V. Chigrinov, Gray scale generation and stabilization in ferroelectric liquid crystal display. *Journal of Display Technology* **3**, 273 (2007).

[16] P. Xu, X. Li, and V. Chigrinov, Double cell achromatic ferroelectric liquid crystal displays using photoalignment technology. *Japanese Journal of Applied Physics* **45**, 200 (2006).

[17] S. Valyukh, I. Valyukh, P. Xu, and V. Chigrinov, Study on birefringent color generation for a reflective ferroelectric liquid crystal display. *Japanese Journal of Applied Physics* **45**, 7819 (2006).

[18] P. Xu, X. Li, A. Muravski, V. Chigrinov, and S. Valyukh, Photoaligned bistable FLC displays with birefringent color switching. *SID'06 Digest*, p. 854 (2006).

[19] E. P. Pozhidaev, G. Hegde, P. Xu, and V. G. Chigrinov, Electrically controlled birefringent colors of smectic C* deformed helix ferroelectric liquid crystal cells. *FLC'07 Program*, O-36 (2007).

[20] M. Omodani, What is Electronic Paper? The expectations. *SID'04 Digest*, p. 128 (2004).

[21] J. W. Doan and A. Khan, Cholesteric liquid crystals for flexible displays. In *Flexible Flat Panel Displays*, Ed. G. P. Crawford, John Wiley & Sons, Ltd, Chichester (2005), 331–354.

[22] J. C. Jones, The Zenithal Bistable Display: from concept to consumer. *Journal of the SID* **16**/1, 143 (2008).

[23] J. Angele, M. Elyaakoibi, F. Guignard, S. Jacquier, S. Joly, P. Martinot-Lagarde, S. Osterman, J. Laffitte, and F. Leblanc, A4 BiNem display with 3.8 mega-pixels. *IDW'07 Digest*, p. 265 (2007).

[24] A. Muravsky, A. Murauski, X. Li, V. Chigrinov, and H.-S. Kwok, Optical rewritable liquid-crystal-alignment technology. *Journal of the SID* **15**/4, 267 (2007).

[25] A. Muravsky, A. Murauski, V. Chigrinov, and H.-S. Kwok, Novel optical rewritable electronic paper. *IDW'07 Digest*, p. 1681 (2007).

[26] A. Muravsky, Optical rewritable photoalignment technology: application in photonics and displays. PhD thesis proposal, HKUST (2008).

[27] A. Zhang, K. T. Chan, M. S. Demokan, V. Chan, P. Chan, H.-S. Kwok, and A. Chan, Integrated liquid crystal optical switch based on total internal reflection. *Applied Physics Letters* **86**, 211108 (2005).

[28] A. Poon, L. Chao, M. Ning, L. S. Lik, D. Tong, and V. Chigrinov, Photonics filters, switches and subsystems for next-generation optical networks. *HKIE Transactions* **11**, 60 (2004).

[29] V. G. Chigrinov, L. Zhou, A. A. Muravsky, and A. W. Poon, Electrically tunable microresonators using photoaligned liquid crystals as cladding layers. US Provisional Patent Application No. 60/794,128, filed on April 24, 2006.

[30] V. G. Chigrinov, Liquid crystal devices for photonics applications. *Proceedings of the SPIE* **6781**, 1M (2007).

7

US Patents Related to Photoalignment of Liquid Crystals

7.1 Introductory Remarks

To the best of our knowledge, a bibliographical review of US patents devoted to photoalignment has never been published despite a number of review articles (see e.g. [7–11] in Chapter 1). Such a review is very important because (i) 70–90% of patent data is never published in other sources; and (ii) web analysis of the patent documents is complicated, since a 'person of ordinary skills' has to make a selection from the enormous amount of a weakly structured information. This chapter is a bibliography of US patents (December 2007 included) on materials and processes for the photoalignment of liquid crystals, liquid crystal displays, optical liquid crystal devices and anisotropic optical elements, and other non-display applications. We restricted our attention to US patents only, since the majority of European, Asian, or other world patents are duplicated in the USA.

Photoalignment of Liquid Crystalline Materials: Physics and Applications
V. Chigrinov, V. Kozenkov and H.-S. Kwok

The bibliography has the following structure: number of patent (family of patents), inventors, assignee, title of patent, international class and current US class, date of patent, and original abstract.

Following the bibliography, Table 7.1 provides the numbers of patents in accordance with the classification described in our book. Table 7.2 includes the names of companies and institutes which are patent owners as well as the numbers of the corresponding patents. A conclusion briefly analyzes the patent data presented.

Table 7.1 Classification of US patents on photoalignment methods of forming the orienting layers of organic substances and devices based on them. (The key words for the classification are in the left column and the numbers of the patents in the right)

1. Polarization-sensitive (photoanisotropic) materials and compositions	
1.1 Photosensitive structures based on photochemical active compounds in solid state	
1.1.1 Polyimides and polyamines	5,612,450; 5,731,405; 5,756,649; 5,807,498; 5,817,743; 5,856,430; 5,856,431; 5,929,201; 5,958,292; 5,958,293; 5,965,691; 5,976,640; 6,001,277; 6,043,337; 6,048,928; 6,060,581; 6,066,696; 6,067,140; 6,084,057; 6,091,471; 6,103,322; 6,107,427; 6,139,926; 6,194,039; 6,200,655; 6,218,501; 6,242,061; 6,268,897; 6,295,111; 6,313,348; 6,335,409; 6,348,245; 6,380,432; 6,451,960; 6,469,763; 6,491,988; 6,552,161; 6,713,135; 6,749,895; 6,770,335; 6,793,987; 6,991,834; 7,005,165; 7,075,607; 7,083,833
1.1.2 Polyvinyl-cinnamates	5,112,881; 5,223,356; 5,290,824; 5,389,698; 5,464,669; 5,484,821; 5,532,320; 5,538,823; 5,539,074; 5,602,661; 5,705,096; 5,767,994; 5,824,377; 5,838,407; 5,859,682; 5,998,563; 6,160,597; 6,215,539; 6,300,991; 6,369,869; 6,383,579; 6,582,784; 6,608,661; 6,717,644; 7,018,687
1.1.3 Silane-type compounds and polymers	5,133,895; 5,515,190; 5,824,377; 5,859,682; 5,909,265; 5,965,761; 5,998,563; 6,091,471; 6,268,897; 6,277,502; 6,295,111; 6,469,763; 7,075,607
1.1.4. Photocrosslinkable polymeric systems	5,112,881; 5,133,895; 5,223,356; 5,290,824; 5,484,821; 5,515,190; 5,532,320; 6,277,502
1.1.5 Chemosorption in monomolecular layers	5,133,895; 5,515,190; 5,578,351; 5,948,316; 6,054,190; 6,284,197; 6,368,681; 6,852,285; 6,858,423
1.1.6 Photocrosslinkable low molecular weight compositions	5,073,294; 6,610,462

Table 7.1 (*continued*)

1.1.7 Polyfunctional azide compounds	5,846,451
1.1.8 Photodimerizable coumarin	6,201,087
1.1.9 Photoreactive liquidcrystalline materials	5,262,882
1.1.10 Other photosensitive compositions	5,838,407; 5,925,423; 5,928,561; 6,144,428; 6,215,539; 6,224,788; 6,300,991; 6,569,972; 6,597,422; 6,608,661; 6,610,462; 6,620,920; 6,627,269; 6,731,362; 6,733,958; 6,764,724; 6,784,231; 6,793,987; 6,797,096; 7,014,892; 7,060,332; 7,074,344; 7,083,833; 7,220,467
1.2. Photosensitive structures on basis of photochemical stable compound in solid state	
1.2.1 Azo dye derivatives	6,582,776; 6,630,289
2 Well-ordered molecular structures	
2.1 Structures based on liquid crystals	
2.1.1 Thermotropic liquid crystals	4,974,941; 5,032,009; 5,389,698; 5,464,669; 5,767,994; 5,784,139; 5,807,498; 5,838,407; 5,958,292; 5,958,293; 5,965,691; 6,200,655; 6,215,539; 6,300,991; 6,608,661
2.1.2 Lyotropic liquid crystals	5,706,131; 6,630,289
2.1.3 PDLC composites	6,083,575
2.1.4 Phase-separated composites	5,949,508; 6,203,866; 6,395,352; 6,939,587
2.2 Structures based on non-liquid crystals anisotropic compounds	
2.2.1 Luminescence compounds	6,243,151; 6,445,431; 6,501,520; 6,594,062; 6,830,831; 7,081,307; 7,118,787
2.2.2 Organic semiconducting compounds	6,737,303
3 Optical devices based on photoalignment layers	
3.1 Exposure by polarized light	5,889,571; 5,936,691; 6,061,138; 6,169,591; 6,292,296; 6,307,609; 6,312,769; 6,312,875; 6,407,894; 6,475,705; 6,512,564; 6,639,720; 6,751,003; 6,781,656; 6,791,749; 6,844,913; 6,874,899; 6,914,708; 6,927,823; 6,943,930; 6,988,811; 7,016,112; 7,016,113; 7,061,679; 7,133,099

(*continued overleaf*)

Table 7.1 (*continued*)

3.2 Exposure by anon-polarized light	6,646,703
3.3 Oblique exposure	5,889,571; 5,909,265; 6,191,836; 6,226,066; 6,414,734; 6,417,905; 6,433,850; 6,462,797; 6,633,355; 6,721,025; 6,879,363; 7,145,618
4 Structure and type of liquid crystal display devices	
4.1 Controlling the liquid crystal pretilt angle	5,604,615; 5,623,354; 5,731,405; 5,817,743; 5,856,430; 5,856,431; 5,929,201; 6,222,601; 6,226,066; 6,292,244; 6,300,993; 6,377,326; 6,414,734; 6,433,850; 6,583,835; 6,633,355; 6,879,363; 7,145,618
4.2 Photoalignment after chemical-mechanical polishing	6,020,249; 6,465,897
4.3 Multi-domain LCD	5,453,862; 5,478,682; 5,479,282; 5,576,862; 5,623,354; 5,824,377; 5,853,818; 5,909,265; 5,953,091; 5,982,466, 5,998,563; 6,067,140; 6,100,953; 6,169,591; 6,292,244; 6,300,993; 6,312,875; 6,335,776; 6,356,335; 6,377,326; 6,449,025; 6,466,288; 6,473,142; 6,475,705; 6,479,218; 6,525,794; 6,593,986; 6,593,989; 6,654,090; 6,661,489; 6,665,035; 6,784,963; 6,787,292; 6,825,906
4.4 LCD devices	5,729,319; 5,953,584; 6,064,451; 6,091,471; 6,184,959; 6,268,897; 6,295,111; 6,383,579; 6,399,165; 6,459,463; 6,466,286; 6,469,763; 6,504,591; 6,549,258; 6,572,939; 6,582,784; 6,583,835; 6,590,627; 6,671,021; 6,812,985; 6,882,391; 6,958,799; 7,018,687; 7,075,607; 7,088,412; 7,244,627; 7,298,429
4.5 In-plane switching mode LCD	6,184,961; 6,259,502; 6,281,959
4.6 Active matrix LCD	5,824,377; 5,928,733; 5,949,509; 5,998,563; 6,242,060; 6,756,089
5 Anisotropic optical elements and devices	
5.1 Optical compensation films	5,995,184; 6,061,113; 6,461,694; 6,582,775; 6,686,980; 6,822,713
5.2 Optical phase retardation films	5,568,294; 5,786,041; 5,807,498; 5,846,451; 5,958,292; 5,958,293; 5,965,691; 6,046,787; 6,128,059; 6,144,428; 6,200,655

Table 7.1 (*continued*)

5.3 Polarizing elements	5,479,282; 5,706,131; 5,751,389; 6,046,849; 6,055,013; 6,055,103; 6,133,973; 6,377,295; 6,404,472; 6,437,915; 6,496,239; 6,624,863; 6,630,289
5.4 Patterned optical elements	5,903,330; 5,926,241; 6,128,058; 6,144,428; 6,630,289; 6,897,926; 6,900,866; 6,909,473; 6,954,245; 6,982,773; 7,044,600; 7,097,303; 7,097,304; 7,101,043; 7,221,420; 7,023,512; 7,061,561; 7,184,115; 7,256,921; 7,286,275
5.5 Optical storage medium	5,846,452; 6,630,289
5.6 Optical security elements	6,496,287; 6,806,930

Table 7.2 Assignees of US patents on the photoalignment method of forming the orienting layers of organic compositions

N	Assignee	Country	Patent no.	Total no. of patents
1	Alliant Techsystems Inc.	USA	5,731,405; 5,807,498; 5,817,743; 5,846,452; 5,856,430; 5,856,431; 5,929,201; 5,958,292; 5,958,293; 5,965,691; 6,200,655	11
2	Eastman Kodak Co.	USA	6,582,775; 6,751,003; 6,844,913; 6,874,899; 6,914,708; 6,943,930; 6,988,811	7
3	E. I. du Pont de Nemours and Company	USA	6,139,926; 6,348,245	2
4	Elsicon Inc.	USA	5,731,405; 5,807,498; 5,817,743; 5,856,430; 5,856,431; 5,929,201; 5,958,292; 5,958,293; 5,965,691; 6,043,337; 6,061,138; 6,084,057; 6,103,322; 6,194,039; 6,200,655; 6,242,061; 6,307,609; 6,313,348; 6,380,432; 6,407,894; 6,451,960; 6,491,988; 6,552,161; 6,713,135; 6,991,834; 7,005,165	26
5	Geo Centers Inc.	USA	5,578,351; 5,948,316	2

(*continued overleaf*)

Table 7.2 (*continued*)

N	Assignee	Country	Patent no.	Total no. of patents
6	Hercules Inc.	USA	4,974,941; 5,032,009; 5,073,294	3
7	IBM Co.	USA	5,623,354; 5,786,041	2
8	Kent State University	USA	5,936,691; 5,949,508; 6,822,713; 6,927,823; 6,939,587	5
9	Moxtek Inc.	USA	6,897,926; 6,900,866; 6,909,473; 6,954,245; 6,982,773; 7,023,512; 7,061,561; 7,184,115; 7,221,420	9
10	OIS Optical Imaging Systems, Inc.	USA	5,953,091	1
11	PPG Industry Ohio, Inc.	USA	7,044,600; 7,097,303; 7,097,304; 7,101,043	4
12	The Regents of the University of California	USA	6,284,197; 6,852,285; 6,858,423	3
13	Rockwell Science Center LLC	USA	5,995,184	1
14	Transition Optical, Inc.	USA	7,256,921; 7,286,275	2
15	University of Massachusetts Lowell	USA	5,112,881; 5,223,356; 5,290,824; 5,484,821; 5,532,320; 6,153,272	6
16	Alps Electric Co., Ltd	Japan	5,705,096	1
17	Cheil Industries Inc.	Japan	6,218,501	1
18	Chisso Corporation	Japan	5,625,475; 5,705,096; 5,798,810	3
19	Dainippon Ink and Chemicals, Inc.	Japan	6,582,776; 6,733,958; 6,784,231	3
20	Fuji Xerox Co. Ltd	Japan	6,083,575	1

Table 7.2 (*continued*)

N	Assignee	Country	Patent no.	Total no. of patents
21	Fujitsu Display Technologies Co.	Japan	6,583,835	1
22	Fujitsu Limited	Japan	6,203,866; 6,395,352; 6,512,564; 6,781,656; 7,133,099	5
23	Fuji Photo Film Co., Ltd	Japan	6,061,113; 6,461,694; 6,686,980	3
24	Hamamatsu Photonics K.K.	Japan	5,625,475; 5,798,810	2
25	Hitachi Ltd	Japan	5,928,733; 5,949,509; 6,001,277; 6,242,060; 6,504,591; 6,590,627; 6,756,089; 6,958,799; 7,088,412	9
26	Iinuma Mfg Co., Ltd	Japan	5,625,475; 5,798,810	2
27	Japan Science and Technology Agency	Japan	7,298,429	1
28	Japan Synthetic Rubber Co. Ltd	Japan	5,478,682; 5,612,450; 5,756,649	3
29	JSR Corporation	Japan	6,224,788; 6,312,769; 7,074,344	3
30	Matsushita Electric Industrial Co. Ltd	Japan	5,133,895; 5,515,190; 6,054,190; 6,368,681	4
31	Nippon Kayaku K.K.	Japan	5,706,131; 6,243,151; 6,445,431; 6,501,520	4
32	Nitto Denko Co.	Japan	5,846,451	1
33	Sharp KK	Japan	5,604,615; 5,729,319; 5,756,649; 6,046,787; 6,046,849; 6,055,013; 6,055,103; 6,128,058; 6,128,059; 6,184,959; 6,377,295; 6,437,915; 6,624,863	13

(*continued overleaf*)

Table 7.2 (*continued*)

N	Assignee	Country	Patent no.	Total no. of patents
34	Simitomo Chemical Co.	Japan	5,751,389; 6,133,973; 6,404,472	3
35	Sony Corporation	Japan	6,897,926; 6,900,866; 6,909,473; 6,954,245; 6,982,773; 7,023,512; 7,061,561; 7,184,115; 7,221,420	9
36	Stanley Electric Ltd	Japan	5,453,862; 5,479,282; 5,576,862	3
37	Gold Star Co. Ltd	Korea	5,464,669; 5,538,823; 5,767,994	3
38	Korea Advanced Institute of Science and Technology	Korea	6,731,362; 6,749,895	2
39	LG Electronics Inc.	Korea	5,824,377; 5,853,818; 5,859,682; 5,882,238; 5,889,571; 5,909,265; 5,953,584; 5,982,466; 5,998,563; 6,064,451; 6,067,140; 6,091,471; 6,169,591; 6,184,961; 6,222,601; 6,259,502; 6,268,897; 6,295,111; 6,312,875; 6,384,888; 6,466,291; 6,469,763; 6,475,705; 6,593,986; 6,593,989; 6,661,489; 6,741,312; 6,812,985; 6,853,429; 7,075,607	30
40	LG LCD Inc.	Korea	6,466,288	1
41	LG. Philips LCD Co., Ltd	Korea	6,100,953; 6,191,836; 6,226,066; 6,281,959; 6,292,296; 6,297,866; 6,335,776; 6,356,335; 6,383,579; 6,396,558; 6,399,165; 6,414,734; 6,417,905; 6,433,850; 6,445,435; 6,449,025; 6,459,463; 6,462,797; 6,466,286; 6,473,142; 6,479,218; 6,525,794; 6,549,258; 6,572,939; 6,582,784; 6,627,269; 6,628,362; 6,633,355; 6,639,720; 6,654,090; 6,665,035; 6,671,021; 6,697,140; 6,721,025; 6,764,724; 6,770,335;	55

Table 7.2 (*continued*)

N	Assignee	Country	Patent no.	Total no. of patents
			6,784,963; 6,787,292; 6,793,987; 6,797,096; 6,798,482; 6,822,717; 6,825,906; 6,879,363; 6,882,391; 7,014,892; 7,016,112; 7,016,113; 7,018,687; 7,060,332; 7,061,679; 7,083,833; 7,145,618; 7,220,467; 7,244,627	
42	Samsung Display Devices Co., Ltd	Korea	5,568,294; 5,925,423; 5,976,640; 6,048,928; 6,060,581; 6,066,696; 6,218,501; 6,569,972	8
43	Hoffmann-La Roche Inc.	USA	5,389,698; 5,539,074; 5,602,661; 6,160,597; 6,369,869; 6,717,644	6
44	Rolic AG	Switzerland	5,539,074; 5,602,661; 5,784,139; 5,838,407; 5,903,330; 5,965,761; 6,107,427; 6,144,428; 6,160,597; 6,201,087; 6,215,539; 6,277,502 6,300,991; 6,335,409; 6,369,869; 6,496,239; 6,597,422; 6,608,661; 6,610,462; 6,646,703 6,717,644	21
45	Hong Kong University of Science and Technology	Hong Kong	6,582,776; 6,630,289	2
46	Industrial Technology Research Institute	Taiwan	6,292,244; 6,300,993; 6,377,326; 6,737,303	4
47	Taiwan Semiconductor Manufacturing Co.	Taiwan	6,020,249; 6,465,897	2
48	NIOPiC Moscow Research and Production Association	Russia	5,389,698	1

(*continued overleaf*)

Table 7.2 *(continued)*

N	Assignee	Country	Patent no.	Total no. of patents
49	The Secretary of State for Defence in Her Britannic Majesty's Government of the United Kingdom of Great Britain and Northern Ireland	UK	5,928,561	1
50	University of Hull	UK	6,830,831; 7,081,307; 7,118,787	3
51	Bayer Aktiengesellschaft	Germany	6,046,290; 6, 620,920	2
52	Merck Patent GmbH	Germany	6,139,926; 6,348,245	2
53	Philips Corporation	USA	5,262,882	1
54	Rockwell International Corporation	USA	5,926,241	1

We hope that our patent review will be useful not only to a wide range of engineers, scientists, and managers who are willing to develop new LCDs or other LC devices and optical components, but also to researchers in other practically important fields, where the formation of new, highly ordered structures of organic molecules is highly desirable. These fields may cover the ordering of thin semiconductor layers, organic films for light-emitting diodes (OLEDs), solar cells, optical data storage, and holographic memory. New and highly efficient photovoltaic, optoelectronic, and photonic devices are envisaged. Such novel structures of organic molecules may exhibit certain physical properties that are similar to those of aligned LC layers. Our patent review will simplify a search for the new photoalignment materials for these applications.

7.2 List of Patents: Patent Classification

1. **US 4,974,941 (US 5,032,009)**, W. M. Gibbons, S. T. Sun, and B. J. Swetlin, (**Hercules Inc.**), *Process of aligning and realigning LC media*, C09K 19/60 (349/24) (December 1990).

   A process of aligning or realigning a liquid crystal medium adjacent to a substrate which process comprises (1) exposing anisotropic ally absorbing molecules disposed on the substrate, disposed in the liquid crystal medium or which themselves are the liquid crystal medium to linearly polarized light of a wavelength or wavelengths within the absorption band of the anisotropically absorbing molecules, wherein (a) the exposed anisotropically absorbing molecules induce alignment of the liquid crystal medium at an angle $\pm\theta$ with respect to the direction of the linear polarization of the incident light beam and along a surface of the liquid crystal medium, (b) the liquid crystal medium comprises liquid crystals having a molecular weight of less than 1500, and (c) the anisotropically absorbing molecules are exposed by (i) linearly polarized light transmitted through at least one mask having a pattern, (ii) a beam of linearly polarized light scanned in a pattern or (iii) patterns created by interference of coherent linearly polarized light beams, and (2) terminating exposure, whereby the liquid crystal medium aligned or realigned by the exposure step remains aligned or realigned and has two or more discrete regions having different alignments, and liquid crystal devices prepared by this process, are disclosed.

2. **US 5,073,294**, P. J. Sannon, S. T. Sun, and B. J. Swetlin, (**Hercules Inc.**), *Process of preparing compositions having multiplexed mesogens*, C07C 255/54 (252/299.01) (December 1991).

   A liquid crystal polymer film or fiber comprising polymeric liquid crystals having aligned multiple oriented mesogens and a process for preparing such a polymeric liquid crystal film or fiber comprising aligning in a multi-oriented state a nematic or smectic monomeric mesophase (liquid crystal monomer) and photopolymerizing the mesophase, are disclosed. The film or fiber may be adhered to a substrate, part of a liquid crystal display cell, or a free standing polymeric film or fiber. According to a preferred process, the mesophase is aligned in a multi-oriented state using linearly polarized light.

3. **US 5,133,895 (US 5,515,190)**, K. Ogawa, N. Mino, and H. Tamura, (**Matsushita Electric Industrial Co., Ltd**), *Alignment film for LC and method for production thereof, as well as LCDD utilizing said alignment film and method for production thereof*, G02F1/1337 (252/299.4) (July 1992).

   An alignment film for liquid crystal formed by monomolecular film comprising silane-type compounds with a linear hydrocarbon chain and method for production of said film, as well as liquid crystal display units utilizing said film and method for production of said display unit are disclosed.

4. **US 5,223,356 (US 5,112,881; US 5,290,824; US 5,484,821; US 5,532,320)**, J. Kumar, S. K. Tripathy, B. K. Mandal, and J. C. Huang, (**University of Lowell**), *Photocrosslinked second order nonlinear optical polymers*, G02F 1/35 (430/1) (June 1993).

   A novel photocrosslinkable polymeric system has been developed for processing into films having stable second-order nonlinear optical properties. In the present system, polymers bearing photocrosslinkable chromophores, such as polyvinylcinnamate, are reacted with appropriately designed nonlinear optical molecules with the cinnamate or other photocrosslinkable functionalities for photocrosslinking at one, two or more points. The system can be poled and photocrosslinked in the poled state to yield a material with stable optical nonlinearity and large electro-optic coefficients.

5. **US 5,389,698**, V. Chigrinov, V. Kozenkov, N. Novoseletsky, V. Resetnyak, Y. Reznikov, M. Schadt, and K. Schmitt, (**Hoffmann-La Roche Inc.; NIOPIC Moscow Research and Production Association**), *Process for making photopolymers having varying molecular orientation using light to orient and polymerize*, G02F 1/1337 (522/2) (February 1995).

   An oriented photopolymer is a polymer where photomodification is initiated by irradiation with plane-polarized light and where the molecular configuration has a preferred orientation as a result of linear polarization. These polymers have optically anisotropic and other novel properties.

6. **US 5,262,882**, R. A. M. Hikmet, (**Philips Corp.**), *Liquid crystal device orientation layer of an oriented polymer network containing liquid crystal*, G02F 1/1337 (349/127) (November 1993).

   An orientation layer for a liquid crystal display device is formed from a mixture of at least one reactive liquid-crystalline material and at least one non-reactive liquid-crystalline material. The mixture is brought into contact with a surface of a mould having an orienting effect, as a result of which the molecules in the mixture are oriented. Subsequently, the reactive liquid-crystalline material is made to cure, thereby forming an oriented polymer network in which the non-reactive liquid-crystalline material is dispersed. The orientation layer thus manufactured is very suitable for use in an active display device.

7. **US 5,464,669 (US 5,767,994)**, D. S. Kang, W. S. Park, H. H. Shin, S. B. Kwon, T. Y. Marusii, Y. A. Reznikov, A. I. Khizhnyak, and O. V. Yaroschuk, (**Gold Star Co., Ltd**), *Method for forming an orientation film of photopolymer in a liquid crystal display,* G02F 1/13 (427/558) (November 1995).

   A method for the formation of orientation film of a liquid crystal display. The method comprises the steps of forming polyvinyl-4-fluorocinnamate (hereinafter 'PVCN-F') film on the surfaces of two opposite substrates and irradiating the two PVCN-F films formed with linearly polarized UV lights having different energy. The method in accordance with the present invention is capable of not only providing the pretilt angle to a fabricated LCD but also adjusting it by illuminating two substrates with linearly polarized UV beams having different energies from each other, respectively. In addition, the LCD fabricated by the method according to the present invention requires a much lower driving voltage, as compared with the conventional LCD having an orientation film of planar structure. Furthermore, phase distortion and light scattering phenomena do not occur in the LCD, so that display characteristics such as contrast and the like can be improved. The orientation film is formed of polyvinyl-4-fluorocinnamate which is highly resistant to heat, so that the LCD according to the present invention has an excellent thermostability.

8. **US 5,478,682**, M. Nishikawa, M. Endoh, Y. Tsuda, and N. Bessho, (**Japan Synthetic Rubber Co., Ltd**), *Method for domain-dividing liquid crystal alignment film and liquid crystal device using domain-divided alignment film*, G02F 1/13 (430/20) (December 1995).

   A method for domain-dividing a liquid crystal alignment film, which comprises (1) applying a radiation-sensitive resin composition onto a liquid crystal alignment film aligned in one direction and formed on a substrate, (2) exposing the radiation-sensitive resin composition through a predetermined pattern, (3) carrying out a development with an aqueous solution containing 0.01 to 1.5% by weight of an alkaline compound to partially protect the liquid crystal alignment film, (4) aligning an exposed portion of the liquid crystal alignment film in another direction and (5) removing the radiation-sensitive resin composition; and a liquid crystal device having a liquid crystal display device with the domain-divided liquid crystal alignment film.

9. **US 5,479,282**, Y. Toko and T. Sugiyama, (**Stanley Electric Co, Ltd**), *Liquid crystal display of multi-domain structure*, G02F 1/1337 (359/75) (December 1995).

   No rubbing treatment is done on a pair of transparent substrates. A liquid crystal layer is sandwiched between these transparent substrates to exhibit a multi-domain structure wherein liquid crystal molecules are oriented to various directions at the interface between the liquid crystal layer and a transparent substrate so that the liquid crystal molecules can be considered to be orientated at every direction at an equal possibility with respect to the directions in a plane parallel to the transparent substrate. The transparent substrates constitute a liquid crystal display cell having a reference direction such as an observation direction. A pair of polarizers are disposed outside the transparent substrates to have a transmission or an absorption axis aligned with the observation direction.

10. **US 5,538,823**, W. S. Park, H. H. Shin, S. B. Kwon, A. Dyadusha, T. Marusii, Y. Rezinkov, A. Khizhnyak, O. Yaroschuk, A. Kolomeytsev, and I. Gerus, (**Gold Star Co., Ltd**), *Liquid crystal device utilizing thermostable polymeric material,* G02F1/1337 (430/20) (July 1996).

    A thermostable orientation material for liquid crystal alignment including a photopolymerized polyvinyl fluorocinnamate.

11. **US 5,539,074 (US 5,602,661; US 6,160,597; US 6,369,869; US 6,717,644)**, R. P. Herr, S. Kelly, M. Schadt, K. Schmitt, and A. Schuster, (**Hoffman-La Roche Inc.**), *Linear and cyclic polymers or oligomers having a photoreactive ethene group*, C08F 246/00 (526/245) (July 1996).

    The invention is concerned with linear and cyclic polymers or oligomers having a photoreactive ethene group, which can undergo photochemical isomerization/dimerization. The compounds are used as an orientating layer for liquid crystals. An optical component includes an anisotropic layer of cross-linked liquid crystal monomers with varying local orientation of the liquid crystal molecules. The liquid crystal layer is in contact with an orientation layer comprising a photo-orientable polymer network (PPN). A method of making includes orienting the liquid crystal monomers by the interaction with the PPN layer and subsequently fixing the molecules by cross-linking.

12. **US 5,568,294**, J. C. Lee, (**Samsung Display Devices Co., Ltd**), *Liquid crystal display having a polymer functioning as an orientation layer and a retarder,* G02F 1/1337 (349/117) (October 1996).

    A liquid crystal display comprises a pair of substrates, transparent electrodes respectively formed thereon and a liquid crystal material layer inserted between the electrodes, and is characterized in that a liquid crystalline polymer orientation layer is formed on at least one of the liquid crystal material layers, and the liquid crystalline polymer orientation layer functions as an optical phase retardation film. The phase retardation of the light transmitting liquid crystal is compensated by the liquid crystalline polymer orientation layer, which enhances contrast. The liquid crystalline polymer layer can also be used as an optical phase retardation film.

13. **US 5,576,862 (US 5,453,862)**, T. Sugiyama, Y. Toko, S. Kobayashi, and Y. Iimura, (**Stanley Electric Co. Ltd**), *Positive orientations of liquid crystal molecules in a multi-domain liquid crystal display cell*, G02F 1/1337 (349/124) (November 1996).

    A liquid crystal display cell comprising a pair of substrates, a liquid crystal material sandwiched between the pair of substrates, and an optical polarization memory film provided on one or both of the substrates. The optical polarization memory film is exposed by polarized lights to produce a positive orientation means including multiple micro domains. In the micro domains, orientation direction is uniform in each domain and substantially isotropic as a whole. The visual angular dependency is eliminated. No rubbing treatment is done, and problems caused by rubbing can be solved.

14. **US 5,578,351 (US 5,948,316)**, R. Shashidhar, B. Peek, B. R. Ratna, J. M. Calvert, J. M. Schnur, M. S. Chen, and R. J. Crawford, (**Geo Centers Inc.**), *Liquid crystal composition and alignment layer,* G02F 1/1337 (428/1.23) (November 1996).

    A surface for the alignment of liquid crystals containing directionally-linked groups.

15. **US 5,604,615**, H. Iwagoe and S. Mizushima, (**Sharp KK**), *Liquid crystal display device and methods for producing same with alignment layer having new bond formation or bond cleavage reaction of molecular chains by light irradiation*, G02F1/1337 (349/124) (February 1997).

    A liquid crystal display device includes: a pair of substrates; a pair of electrodes formed on the pair of substrates; a pair of alignment films formed on the pair of electrodes; and a liquid crystal layer sandwiched between the pair of alignment films, wherein at least one of the alignment films have a property of aligning liquid crystal molecules of the liquid crystal layer with a pretilt angle, the property being provided by a new bond formation reaction or bond cleavage reaction of molecular chains caused by irradiating at least one alignment film with light.

16. **US 5,612,450 (US 5,756,649)**, S. Mizushima N. Watanabe, H. Iwagoe, S. Makino, S. Kawamura, Y. Tsuda, and N. Bessho, (**Japan Synthetic Rubber Co., Ltd; Sharp KK**), *Liquid crystal aligning agent and liquid crystal display device*, G02F1/1337 (528/353) (March 1997).

    A liquid crystal aligning agent comprising a polyamic acid containing an aliphatic and/or alicyclic hydrocarbon group and a polyimide containing an aliphatic and/or alicyclic hydrocarbon group; and a liquid crystal display device using the liquid crystal aligning agent. This liquid crystal aligning agent gives a liquid crystal aligning film which has good liquid crystal aligning property and in which pretilt angle can be changed by radiation with a small energy and which is suitable for domain-divided alignment type liquid crystal display having a wide view angle.

17. **US 5,623,354**, S. C. Lien, R. A. John, and M. Angelopoulos, (**IBM Co.**), *Liquid crystal display with multi-domains*, G02F1/1337 (349/124) (April 1997).

    A liquid crystal display with an array of pixels comprises a liquid crystal layer having transparent electrodes on either side thereof with an alignment layer disposed on each of the electrodes and in contact with opposite surfaces of the liquid crystal layer. The alignment layers are capable of inducing a predetermined pre-tilt angle on the molecules of the liquid crystal. The pre-tilt angle inducing ability of at least one of the alignment layers, which is formed with only one alignment direction, has been selectively altered so that pre-tilt angle inducing ability in the altered area differs in the magnitude of the angle from the pre-tilt angle inducing ability of the unaltered area. With the liquid crystal molecules having at least two different pre-tilt angles, more than one domain is created when voltage is applied across the electrodes. The change in the pre-tilt angle inducing ability in the altered areas is achieved by exposure to a beam, such as ultraviolet light, of sufficient energy. In the preferred embodiment, the display is an active type with twisted nematic liquid crystals.

18. **US 5,625,475 (US 5,798,810)**, S. Tanioka, S. Murata, M. Kono, and M. Hirano, (**Chisso Corp.; Iinuma MFG 8Co. Ltd; Hamamaysu Photonics KK**), *Method for treating an aligning film for a liquid crystal display element*, G02F1/1337 (349/123) (April 1997).

    A method for treating an aligning film for a liquid crystal display element wherein an aligning film consisting of an organic high molecular compound formed on a transparent substrate with a transparent electrode is subjected to a robbing treatment and thereafter said aligning film is irradiated with soft X-rays in a gas. A method for preparing a liquid crystal display element comprising the steps of arranging a transparent electrode on a pair of transparent substrates, forming an aligning film consisting of an organic high molecular compound on the said transparent electrode, treating said aligning film by a rubbing treatment, assembling the substrates to face each other with a spacer disposed there between, and then applying and sealing to a substrate a liquid crystal material wherein the aligning film is subjected to a robbing treatment and thereafter said aligning film is irradiated with soft X-rays in gas.

19. **US 5,705,096**, M. Kano, Y. Ishitaka, Y. Sato, K. Yoshii, S. Sugimori, and T. Katoh, (**Alps Electric Co., Ltd; Chisso Co.**), *UV crosslinking compound, alignment film for LCD component and LCD component*, C08F 18/20 (252/299.4) (January 1998).

    The UV crosslinking compound of the present invention is characterized in having the following chemical formula

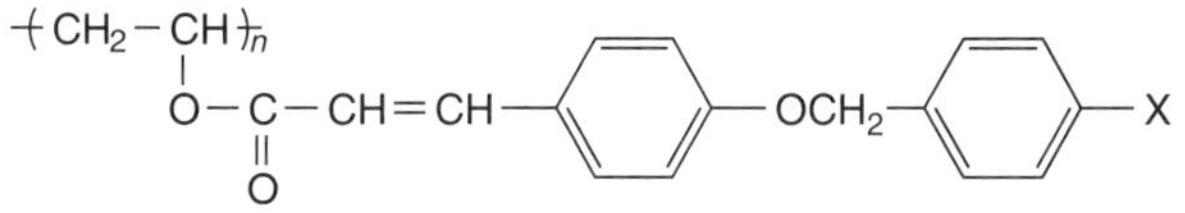

$$-(CH_2-CH)_n- \quad | \quad O-C(=O)-CH=CH-C_6H_4-OCH_2-C_6H_4-X$$

and has high photosensitivity. Accordingly, the crosslinking reaction can be accomplished within a short period of time. Thus, if this compound is employed in the alignment film for a LCD component, it becomes possible to carry out a sufficient alignment treatment with UV irradiation over a short period of time. Accordingly, production time can be shortened and production costs reduced.

20. **US 5,706,131**, K. Ichimura, N. Ishizuki, and J. Toda, (**Nippon Kayaku Kabushiki Kaisha**), *Polarizing element, polarizing plate, and process for production thereof*, G02B 5/30 (359/490) (January 1998).

    A polarizing element or polarizing plate comprising a layer having photoactive molecules and a layer containing dichroic molecules formed in contact with said layer, which can easily be produced without a stretching procedure so as to have a complicated pattern, a curved surface or a large area; and a process for producing said polarizing element or polarizing plate which is characterized by irradiating a layer having photoactive molecules on a substrate with linear polarized light, and then forming a dichroic molecular layer on the irradiated layer.

21. **US 5,729,319**, I. Inou, Y. Shirai, and M. Shiomi, (**Sharp KK**), *Liquid crystal display device and method for fabricating the same*, G02F1/13 (349/156) (March 1998).

    A liquid crystal display device includes a pair of substrates opposed to each other, polymeric walls patterned in a predetermined pattern, and a liquid crystal layer at least a part of which is surrounded by the polymeric wall, and the polymeric wall and the liquid crystal layer are interposed between the substrates, wherein the polymeric wall has a predetermined rotatory polarization corresponding to an alignment regulating force of the substrates.

22. **US 5,731,405 (US 5,817,743; US 5,85,6430; US 5,856,431; US 5,929,201)**, W. M. Gibbons, P. J. Shannon, and S. T. Sun, (**Alliant Techsystems Inc.**), *Process and materials for inducing pre-tilt in liquid crystals and liquid crystal displays*, G0F1/1337 (528/353) (March 1998).

    A process for inducing pre-tilt in alignment of a liquid crystal medium comprising exposing at least one optical alignment layer, comprising anisotropically absorbing molecules and hydrophobic moieties, to polarized light; the polarized light having a wavelength within the absorption band of said anisotropically absorbing molecules; wherein the exposed anisotropically absorbing molecules induce alignment of the liquid crystal medium at an angle $\pm\theta$ with respect to the direction of the polarization of the incident light beam and along the surface of the optical alignment layer, and induce a pre-tilt at an angle PHI with respect to the surface of the optical alignment layer and applying a liquid crystal medium to said optical alignment layer, is described. The invention also is directed to liquid crystal display elements made by the process of the invention and to novel polyimide compositions that are useful as optical alignment layers in the invention.

23. **US 5,751,389 (US 6,133,973; US 6,404,472)**, A. Andreatta and S. Doi, (**Simitomo Chemical Co.**), *Film containing oriented dye, method of manufacturing the same, and polarizer and LCD unit utilizing the same*, G02F 1/13 (349/97) (May 1998).

Films containing oriented dye, methods of manufacturing such films, and polarizers and liquid crystal display devices utilizing the films are disclosed. The film contains at least one dichroic dye which is uniaxially oriented. A film has a thickness of not less than 1 nanometer and not greater than 5 micrometer and contains the dichroic dye of not less than 1 percent by weight and not greater than 100 percent by weight. The film has a peak of absorption at 400 nm through 800 nm and a dichroic ratio of not less than 25 at the peak of absorption. The film may include a micro-pattern having a minimum width in a film plane of not less than 1 micrometer and not greater than 200 micrometer. In such a case, the dichroic dye has a dichroic ratio of not less than 10 at the peak of absorption. A method of manufacturing such a film according to the invention includes the step of forming a dye-containing film on a substrate having a fluororesin alignment layer on the surface thereof. The dye-containing film includes at least one dichroic dye or a mixture of at least one dichroic dye and at least one polymer compound. In subsequent steps, the dye-containing film may be stripped off from the substrate and transferred to another substrate.

24. **US 5,784,139**, V. Chigrinov, V. Kozenkov, N. Novoseletsky, V. Reshetnyak, Y. Reznikov, M. Schadt, and K. Schmitt, (**Rolic AG**), *Image display device*, H04N 13/00 (349/117) (July 1998).

The image-producing device comprises a liquid crystal cell with a liquid crystal between two plates provided with drive electrodes and wall orientation layer and an optically anisotropic polymeric additional element. The additional element, which can be a separate element or in the form of a coating or a foil stuck directly to the cell, comprises an oriented photopolymer.

25. **US 5,786,041**, A. Takenaka, K. Tajima, H. Takano, S.-C. Lein, and K.-W. Lee, (**IBM Co.**), *Alignment film, a method for producing the alignment film and a liquid crystal display device using the alignment film*, G02F1/1337 (428/1.2) (July 1998).

    An alignment-film in which the pretilt angle is changed by UV irradiation, a material for forming the same, a liquid crystal display device using the alignment film, a method for producing an alignment-film formation material, and a method for producing an alignment film. UV irradiation causes large changes in pretilt angle but few changes in electrical characteristics, and a liquid crystal display device free from deterioration of the liquid crystal, image sticking, or the occurrence of flicker results. A first polymer forms the underlying principal layer and a second polymer forms thereon a surface layer thinner than the principal layer. The first polymer may exhibit a small change in electrical characteristics due to UV radiation and the second polymer may exhibit a large change in pretilt angle due to UV radiation.

26. **US 5,807,498 (US 5,958,292; US 5,958,293; US 5,965,691; US 6,200,655)**, W. M. Gibbons, P. J. Shannon, and S. T. Sun, (**Alliant Techsystems Inc.**), *Process and materials for alignment of liquid crystal and liquid crystal optical element*, C08G73/10 (252/299.4) (September 1998).

    A process for aligning liquid crystals adjacent to a surface of an optical alignment layer comprising: exposing at least one optical alignment layer, comprising anisotropically absorbing molecules, to polarized light; the polarized light having a wavelength within the absorption band of said anisotropically absorbing molecules; wherein the exposed anisotropically absorbing molecules induce alignment of a liquid crystal medium at an angle $\pm\theta$ with respect to the direction of the polarization of the incident light beam and along the surface of the optical alignment layer; and applying a liquid crystal medium to said optical alignment layer; wherein said anisotropically absorbing molecules consist essentially of diaryl ketones, is described. The invention also is directed to a liquid crystal optical storage medium, a liquid crystal display element, and a liquid crystal diffractive optical element made by the process of the invention and to novel polyimide compositions that are useful as optical alignment layers in the process of the invention.

27. **US 5,824,377 (US 5,998,563)**, G. Pirwitz, H. Zaschke, A. Hohmuth, Y. Reznikov, O. Yaroschuk, and I. Gerus, (**LG Electronics Inc.**), *Photosensitive material for orientation of liquid crystal device and liquid crystal device thereof*, C08G 77/00 (428/1.23) (October 1998).

    Photosensitive orientation material providing unidirectional LC alignment to generate a nigh pretilt angle for various LC compounds, and superior thermal stability, and being suitable for mass production, especially for active matrix LC displays, including polysiloxane, and a derivative of a cinnamoyl group.

28. **US 5,838,407 (US 6,215,539; US 6,300,991; US 6,608,661)**, V. Chigrinov, V. Kozenkov, N. Novoseletsky, V. Reshetnyak, Y. Reznikov, M. Schadt, and K. Schmitt, (**Rolic AG**), *Liquid crystal display cells*, G02F1/1337 (349/117) (November 1998).

    In liquid crystal display cells comprising a liquid crystal disposed between two substrate plates provided with drive electrodes, the orientation layer on the inner sides of the plates where the electrodes are disposed comprises an oriented photopolymer layer. This layer is formed by irradiating photopolymer layer. This layer is formed by irradiating photoreactive material with linear polarized light.

29. **US 5,846,451**, S. Nakano, A. Mochizuki, H. Motomura, and K. Izumi, (**Nitto Denko Co.**), *Crosslinking type liquid crystal polymer and oriented crosslinking film thereof*, C09K19/38 (252/299.01) (December 1998).

    A crosslinking type liquid crystal polymer, an oriented crosslinking film thereof, and a method of producing the oriented crosslinking film are disclosed. The crosslinking type liquid crystal polymer comprises a liquid crystal polymer containing a polyfunctional azide compound represented by the following formula (A):

    $$R'\text{—}(N_3)_n \qquad \text{(A)}$$

    wherein R′ represents an organic group and n represents an integer from 1 to 5.

30. **US 5,846,452**, W. M. Gibbons, S. T. Schnelle, P. J. Shannon, and S. T. Sun, (**Alliant Techsystems Inc.**), *Liquid crystal optical storage medium with gray scale*, G11B 7/25 (252/299.4) (December 1998).

    An optical storage medium with gray scale capability comprising: a plurality of facing substrates in series; alignment layers disposed on one or both sides of the substrates to give alignment layer pair(s) wherein at least one of the alignment layers of each of the alignment layer pair(s) is an optical alignment layer, and wherein the optical alignment layer(s) comprise anisotropically absorbing molecules or moieties; liquid crystal layers disposed between the alignment layer pair(s); wherein each of the liquid crystal layer(s) comprises three or more alignment regions having three or more different alignment states; and wherein the different alignment states are controlled by exposure of optical alignment layer(s) with polarized light of a wavelength or wavelengths within the absorption band of the anisotropically absorbing molecules or moieties therein, is described. The invention is also directed to compact disc, erasable compact discs, photographic storage devices, and storage displays.

31. **US 5,853,818 (US 6593986)**, S. B. Kwon, J. H. Kim, O. Yaroschuk, and A. Dyadyusha, (**LG Electronics Inc.**), *Method for manufacturing multi-domain liquid crystal cell*, C08G77/00 (427/510) (December 1998).

    A method for fabricating a multi-domain liquid crystal cell is disclosed, wherein first and second alignment directions are formed in first and second portions of an alignment layer provided on a substrate by selectively subjecting the first and second portions to different energy doses of linearly polarized ultraviolet light. Liquid crystal material is then injected between the one substrate and another substrate and into contact with the alignment layer, thereby obtaining a wide viewing angle in the liquid crystal device.

32. **US 5,859,682**, J. H. Kim, K. H. Yoon, J. W. Woo, M. S. Nam, and Y. J. Choi, (**LG Electronics Inc.**), *Method for manufacturing liquid crystal cell using light*, G02F1/1337 (349/124) (January 1999).

    A twisted nematic liquid crystal display which has a pixel divided into at least one is fabricated by using a UV light irradiated photo-polymer layer provided on a substrate, which includes a polysiloxane based material or polyvinylfluorocinnamate. The alignment direction and pretilt angle direction are adjusted by anchoring energy and flowing effect of liquid crystal material between substrates.

33. **US 5,882,238**, J. H. Kim, K. H. Yoon, J. W. Woon, M. S. Nam, and Y. J. Choi, (**LG Electronics Inc.**), *Method for manufacturing bend-aligned liquid crystal cell using light*, G02F 1/1337 (445/24) (March 1999).

    A bend-aligned liquid crystal cell is manufactured by irradiating first and second alignment layers with first and second polarized ultraviolet lights to impart first and second pretilt angles in the first and second alignment layers, and injecting molecules of liquid crystal material between the first and second alignment layers. The alignment layers include polysiloxane based materials or polyvinylfluorocinnamate in which the pretilt angle of the alignment layer is changeable according to the ultraviolet absorption energy.

34. **US 5,889,571**, J. H. Kim, K. H. Yoon, J. W. Woo, M. S. Nam, Y. J. Choi, and J. H. Jung, (**LG Electronics Inc.**), *Ultraviolet irradiating device for photoalignment process and an irradiating method using the same*, G02F 1/1337 (349/124) (March 1999).

    A photo-irradiating apparatus includes a UV lamp generating the UV light, a lens, a polarizer in which the UV light is linearly polarized, and a substrate exposed by the linearly polarized UV light. The UV lamp, the lens, and the polarizer are arranged in a line. The UV light is irradiated into the alignment layer during the moving of a scan motor to irradiate the whole area of the alignment layer uniformly. In addition, the light is obliquely irradiated into the alignment layer so that the area exposed to the light is enlarged.

35. **US 5,903,330**, J. Fuenfschilling, M. Schadt, M. Ruetschi, and H. Sieberle, (**Rolic AG**), *Optical component with plural orientation layers on the same substrate wherein the surfaces of the orientation layers have different patterns and direction*, G02F 1/1337 (349/129) (May 1999).

    An optical component includes a substrate or cell having two spaced-apart substrates, one or more orientation layers on the substrates and one or more anisotropic layers of cross-linked liquid crystalline monomers or oligomers with locally different orientation of the liquid crystal molecules. The surfaces of the orientation layers adjacent the liquid crystalline layers have orientation patterns with a defined parallel or fan-like line structure in locally limited regions. The average spacing between the lines of the fan-like line structure is not greater than the thickness of the liquid crystal layer and the angle between neighboring lines is not greater than 3 degrees.

36. **US 5,909,265**, J. H. Kim, Y. J. Choi, K. H. Yoon, M. S. Nam, and J. W. Woo, (**LG Electronics Inc.**), *Method of fabricating a multi-domain LCC*, G02F1/1337 (349/129) (June 1999).

    A method of fabricating multi-domain liquid crystal cell includes the steps of providing first and second substrates, the first and second substrates being coated with photoalignment layer, covering the substrate with a mask which has plurality regions having different transmittances, exposing the substrate to vertical light having a first polarization direction, and exposing the substrate to oblique light. The photoalignment materials include polysiloxane-based materials.

37. **US 5,925,423**, K. Y. Han and S. H. Yu, (**Samsung Display Devices Co., Ltd**), *Optical alignment polymer, alignment layer formed using the same and liquid crystal display device having the alignment layer*, C08G 73/00 (358/1.15) (July 1999).

    An optical alignment polymer having an optical aligning functional group simultaneously introduced into its main chain and side chain, an alignment layer formed using the optical alignment polymer and a liquid crystal display device having the alignment layer are provided. The optical alignment polymer has excellent alignment performance and stability against heat or impact.

38. **US 5,926,241**, W. J. Gunning III, (**Rockwell International Corp.**), *Photo-patterned compensator with thin film having optically birefringent and isotropic regions and method of manufacturing for a liquid crystal display*, G02B 5/30 (349/117) (July 1999).

    A method to fabricate a photo-patterned organic compensator for liquid crystal displays, and the resulting compensator structure, are described. One illustrative fabrication method comprises: (1) depositing a thin film of polymerizable liquid crystal material onto one of a display's substrates; (2) orienting the liquid crystal material's director to a specified direction; (3) polymerizing the deposited thin film through an aperture mask; (4) removing the aperture mask; and (5) curing/annealing the thin film layer to yield a planar structure having substantially birefringent and isotropic regions. Complete pixelated compensators may be created by layering two or more such two-region thin films.

39. **US 5,928,561**, G. P. Bryan-Brown, I. C. Sage, W. J. Feast, and K. E. Foster, (**The Secretary of State for Defence in Her Britannic Majesty's

Government of the United Kingdom of Great Britain and Northern Ireland (GB)), *Polymers for liquid crystal alignment*, G02F 1/1337 (252/299.4) (July 1999).

A method of aligning liquid crystal material by bringing into contact with a thin layer of polymer which contains the anthracenyl group, the polymer having previously been treated with plane polarised UV light. A liquid crystal cell is described wherein each of the substrates defining two opposing cell walls are coated on their inner surface with a suitable polymer for carrying out the invention. A typical embodiment of the invention would have repeat unit (I).

40. **US 5,929,201 (US 5,731,405; US 5,817,743; US 5,856,430; US 5,856,431)**, W. M. Gibbons, P. J. Sannon, and S. T. Sun, (**Elsicon Inc.**), *Fluorinated amine products*, C08G 73/00 (528/353) (July 1999).

    The present invention relates to amine compositions and the preparation of polyimides. The polyimides can be used for inducing alignment of a liquid crystal medium with polarized light and liquid crystal display elements.

41. **US 5,936,691**, S. Kumar and J. H. Kim, (**Kent State University**), *Method of preparing alignment layer for use in LCDs using in-situ ultraviolet exposure*, C02F 1/1337 (349/124) (August 1999).

    A non-contacting method of forming an alignment layer on a substrate used in liquid crystal displays which includes the steps of cleaning a substrate surface, disposing a solution having a prepolymer, such as polyamic acid or a resin and a curing agent, and solvent on the substrate surface, evaporating the solvent, and positioning an UV light source proximally near the substrate surface. A linear polarizer is positioned between the UV light source and the substrate surface. UV light is projected through the polarizer onto the substrate surface to simultaneously molecularly align the polymer segments as the prepolymer is polymerized to form an alignment layer on the substrate. Adjusting the direction of polarization and the angle of incidence of the UV light source allows for generation of an alignment layer with a corresponding pre-tilt angle.

42. **US 5,949,508**, S. Kumar and V. Vorflousev, (**Kent State University**), *Phase separated composite organic film and methods for the manufacture thereof*, G02F 1/13 (349/122) (September 1999).

A light modulating cell comprises a pair of substrates, alignment layers disposed on at least one of the substrates and a solution of polymerizable prepolymer and low molecular weight organic material disposed between the pair of substrates. The solution is phase separated and forms a layer of polymeric material and a layer of organic material between the two substrates. An external force may then be applied across the substrates to alter the optical appearance of the layer of organic material from one state to another. A photo-sensitive layer may also be provided in the cell.

43. **US 5,949,509**, M. Ohe, S. Matsuyama, and M. Matsusida, (**Hitachi Ltd**), *Active matrix LCDD method for checking the alignment ability of a photoalignment layer*, G02F 1/1337 (349/123) (September 1999).

    A first substrate has at least scanning signal electrodes, video signal electrodes and pixel electrodes, all of which constitute display pixels, and active elements. A first alignment layer is formed, directly or via an insulating layer, as a top layer on the electrodes constituting the display pixels and the active elements. A second substrate which is bonded to the first substrate with a very small interval provided in between is provided with a second alignment layer that is formed so as to confront the first alignment layer. A liquid crystal layer is provided in a space between the first and second alignment layers. The electrodes constituting the display pixels are formed so as to apply, to the liquid crystal layer, an electric field that is substantially parallel with the surfaces of the first and second substrates and are connected to an external control device for controlling the electric field in accordance with a desired display pattern. Polarizing plates change an optical characteristic of light to be output from the second substrate in accordance with the alignment state of the liquid crystal layer that is produced by the electric field applied to the liquid crystal layer. At least one of the first and second alignment layers has been given alignment ability by illumination with polarized light.

44. **US 5,953,091**, M. R. Jones and L. D. Lovshe, (**OIS Optical Imaging Systems, Inc.**), *Multi-domain LCD and method of making same*, G02F 1/1337 (349/129) (September 1999).

    A multi-domain liquid crystal display includes multiple liquid crystal alignment domains per pixel. In certain embodiments, only a single mechanical buffing step is required per substrate to provide the multiple domains per pixel. A continuous base underlying mechanically buffed alignment layer is provided, and a reactive alignment layer is provided thereon. The reactive alignment layer is doped so as to cause twist of molecules therein. The reactive layer may be photo-polymerizable and patterned so as to form an array of alignment portions and expose/uncover a corresponding array of areas of the underlying buffed alignment layer. Thus, liquid crystal molecules contacting the patterned reactive alignment portions are aligned in one direction while liquid crystal molecules contacting the exposed mechanically buffed polyimide layer are aligned in another direction. Multiple domains are thus provided, with only a single mechanical buffing step being necessary.

45. **US 5,953,584**, K. N. Lim and J. H. Kim, (**LG Electronics Inc.**), *Method of fabricating liquid crystal display device having alignment direction determined*, H01L 21/84 (438/30) (September 1999).

    A method of fabricating a liquid crystal display device having a substrate includes the steps of forming a gate electrode on the substrate, forming a gate insulating layer on the gate electrode, forming a semiconductor layer on the gate insulating layer, forming source/drain electrodes on the semiconductor layer, forming a pixel electrode on the source/drain electrodes including the gate insulating layer, forming a passivation layer on the pixel electrode including the source/drain electrodes, forming a light shielding layer on the passivation layer, forming an alignment layer on the light shielding layer including the passivation layer, and determining an alignment direction by exposing the alignment layer to a light.

46. **US 5,958,292 (US 5,807,498; US 5,958,293; US 5,965691; US 6,200,655)**, W. M. Gibbons, P. J. Shannon, and H. Zheng, (**Elsicon Inc.**), *Materials for inducing alignment of liquid crystals and liquid crystal optical elements*, C08G 73/10 (252/299.4) (September 1999).

    A polyamic acid derived from an amine component comprising a 2-cyano-1,4-phenylene diamine and a family of diaryl ketones is claimed. The polyamic acids are useful in formation of polyimides, also claimed, and for the optical alignment of liquid crystals and the manufacture of liquid crystal optical elements, also claimed.

47. **US 5,958,293 (US 5,807,498; US 5,958,292; US 5,965691; US 6,200,655)**, W. M. Gibbons, P. J. Shannon, and H. Zheng, (**Elsicon Inc.**), *Process for inducing alignment of liquid crystals and liquid crystal optical elements*, C08G 73/10 (252/299.4) (September 1999).

A process for inducing alignment of a liquid crystal medium adjacent to a surface of an optical alignment layer comprising: exposing at least one optical alignment layer to polarized light; the polarized light having a wavelength within the absorption band of said optical alignment layer; wherein the exposed alignment layer induces alignment of the liquid crystal medium at an angle $\pm\theta$ with respect to the direction of the polarization of the incident light beam and along the surface of the optical alignment layer; and applying a liquid crystal medium to the optical alignment layer, wherein the optical alignment layer is a polyimide comprising an amine component having a 2-substituted 1,4-benzenediamine wherein the 2-substituent X1 is an electron withdrawing group having a positive sigma. Also claimed is a liquid crystal display element made by the process.

48. **US 5,965,761**, R. Buchecker, F. Herzog, G. Marck, and A. Schuster, (**Rolic AG**), *Photoactive silane derivatives for liquid crystals*, C09K19/40 (556/440) (October 1999).

    The invention provides novel cross-linkable, photoactive silane derivatives and mixtures with 3-aryl-acrylic acid esters and amides. The present invention also provides orienting layers for liquid crystals, and non-structured and structured optical elements and multi-layer systems based on the silane mixtures and derivatives.

49. **US 5,976,640**, H. S. Yu and B. H. Chae, (**Samsung Display Devices Co., Ltd**), *Optical alignment composition, alignment layer formed using the same and LCD having the alignment layer*, C08G 73/10 (428/1.1) (November 1999).

    An optical alignment composition including polyimide having a cinnamate group at its side chain is disclosed. Also, there are disclosed an alignment layer formed using the composition and a liquid crystal display device (LCD) having the same. The alignment layer has an excellent thermal stability. According to the present invention, since the alignment layer with excellent thermal stability is obtained, the LCD having excellent performance can be manufactured.

50. **US 5,982,466**, Y. J. Choi and J. H. Jung, (**LG Electronics Inc.**), *Method of forming an alignment layer and a liquid crystal cell having same using patterning by light irradiation*, G02F 1/13 (349/124) (November 1999).

    A method of fabricating a liquid crystal cell comprises the steps of providing a substrate whose one surface is coated with an alignment material containing spacers by spin coating method, and patterning an alignment layer by irradiating a light to the selected part of the alignment material.

51. **US 5,995,184**, Y. J. Chung, Z. Zhuang, Z. Li, B. K. Winker, and J. H. Hanamoto, (**Rockwell Science Center LLC**), *Thin film compensators having planar alignment of polymerized liquid crystals at the air interface*, G02B 5/30 (349/118) (November 1999).

    The present invention relates to thin film retardation plates, such as cholesteric and A-plate compensators, for improving the viewing angle and contrast of liquid crystal displays. The A-plate is fabricated using a single substrate onto which a layer of polymerizable reactive mesogens (RMs) is solvent cast. Included in the RMs solution is an additive that migrates to the RM/air interface to lower the surface energy and generate an additive-rich surface layer, which in turn lowers the intrinsic tilt angle of the RMs at the air interface to between 25 degrees and about zero degrees. The solvent is evaporated and the resulting film is polymerized in the liquid crystal phase to permanently orient the liquid crystal. The resulting film may be readily separated from the substrate and shaped into any desired pattern and combined to form novel compensators. In an alternative embodiment, a tilt may be provided by treating the substrate with an oblique alignment layer to form a compensator having a defined splay through the bulk of the thin film.

52. **US 6,001,277**, K. Ichimura, N. Miyadera, Y. Miyadera, Y. Honda, I. Fukuchi, N. Ohta, and P. S. Johar, (**Hitachi Chemical Co., Ltd**), *Liquid-crystal alignment film*, G02F 1/1337 (252/299.4) (December 1999).

    Invention related to a liquid-crystal alignment film that can align liquid-crystal molecules without resort to the rubbing. The liquid-crystal alignment film of the present invention comprises a resin (e.g., a polyimide) containing a photoisomerizable and dichroic structural unit (e.g., a stilbene derivative), and is furnished with the ability to align liquid-crystal molecules when a film formed of the resin is irradiated with linearly polarized light; the ability to align liquid-crystal molecules being held and fixed. The liquid-crystal alignment film of the present invention is used in electric-field driven type liquid-crystal display devices.

53. **US 6,020,249 (US 6,465,897)**, T. Shih, J. Y. Chang, S. M. Jang, and C. H. Yu, (**Taiwan Semiconductor Manufacturing Co.**), *Method for photo alignment after CMP planarization,* G03F 9/00 (438/401) (February 2000).

A method for forming alignment marks are disclosed for performing photoalignment after chemical–mechanical polishing (CMP). A trench is first formed in a silicon substrate and then alignment marks are formed at the bottom of the trench. The aspect ratio of the trench is selected to be so low that the dishing of the CMP pad can be prevented from reaching into the trench to damage the alignment marks therein. A trench structure is also provided whereby the alignment marks can be protected from the abrasive action of the CMP. Steps subsequent to the CMP can therefore proceed unimpeded with the presence of undamaged alignment marks.

54. **US 6,046,787 (US 6,128,059)**, K. Nishiguchi, (**Sharp KK**), *Stereoscopic optical element including birefringent photosensitive film having regions of mutually different prescribed slow axes or fast axes, and an image display device using the same*, G02F 1/13 (349/129) (April 2000).

    An optical element includes a transparent substrate having a main surface, a first birefringent photosensitive film arranged on a first region of the main surface such that a slow axis or a fast axis is aligned with a prescribed first direction, and a second birefringent photosensitive film formed on a second region different from the first region of the main surface such that the slow axis or the fast axis is aligned with a prescribed second direction different from the first direction. The first and second directions are different by 90° ($\pm 20^\circ$, more preferably, $\pm 10^\circ$). An image display device employing the optical element further includes an image forming device such as a liquid crystal display device, and with the optical element arranged on an image display screen, stereoscopic imaging is possible.

55. **US 6,048,928**, H. S. Yu, K. Y. Han, S. H. Yu, Y. K. Jang, B. H. Chae, and J. K. Song, (**Samsung Display Devices Co., Ltd**), *Optical alignment polymer and optical alignment composition having the same*, C08F 122/40 (525/35) (April 2000).

    An optical alignment polymer and an optical alignment composition having the same are provided. The optical alignment polymer has a compound selected from the group consisting of poly(maleimide), poly(vinyl acetal), poly(phenylene oxide), poly(maleimide styrene) and their derivatives in its main chain, and a photosensitive group at its side chain.

56. **US 6,054,190**, K. Ogawa, T. Nomura, and T. Ootake, (**Matsushita Electric Industrial Co., Ltd**), *Method for producing an alignment chemisorption monomolecular film*, G02F 1/13 (427/510) (April 2000).

A method for producing an alignment chemisorption monomolecular film having high efficiency of washing solution and being excellent in alignment property of molecules constituting the alignment monomolecular film. The method comprises the step of forming a monomolecular film by chemisorption on a hydrophilic surface of a base material by bringing the hydrophilic surface into contact with a silane-based surfactant having a carbon chain or a siloxane bond chain so as to cause a chemical reaction between them, thereby bonding one end of the surfactant molecules to the hydrophilic surface, orienting the base material in a predetermined orientation, and then vapor washing the base material in vapor of an organic solvent, and aligning molecules constituting the monomolecular film by the flow of condensation of the organic solvent formed on the surface of the base material during the vapor washing.

57. **US 6,055,103 (US 6,046,849; US 6,055,013; US 6,055,103; US 6,377,295; US 6,437,915; US 6,624,863)**, G. J. Woodgate, A. M. S. Jacobs, J. Harrold, R. R. Moseley, and D. Ezra, (**Sharp KK**), *Passive polarisation modulating optical element and method of making such an element*, G02B 27/00 (359/494) (April 2000).

    A passive polarisation modulating optical element comprises a layer of birefringent material. The layer has substantially fixed birefringence and comprises retarder regions forming a regular pattern, for instance to act as a parallax barrier for a 3D display. The retarders have optic axes aligned in different directions from each other. The element may be associated with a polariser, for instance an output polariser of a liquid crystal device, with the polarising direction of the polariser being parallel to the optic axis of the retarders. Thus, the retarders have no effect on the intensity of light passing through the element whereas the retarders act as half waveplates and rotate the polarisation vector of light passing therethrough, for instance by 90 degrees.

58. **US 6,060,581**, H. S. Yu and Y. K. Jang, (**Samsung Display Devices Co. Ltd**), *Optical alignment polymer, optical composition, alignment layer formed using the same and LCD having the alignment layer*, G03F 7/00 (528/353) (May 2000).

    An optical alignment composition including self-photosensitive polyimide having a benzophenone moiety and an active hydrogen moiety, and an LCD having the alignment layer formed of the optical alignment composition are provided. Since the alignment layer with excellent thermal stability and improved pretilt angle is obtained, the LCD having excellent performance can be manufactured.

59. **US 6,061,113**, K. Kawata, (**Fuji Photo Film Co., Ltd**), *Optical compensatory sheet having an orientation layer activated by irradiation of light, process and preparation of the same and liquid crystal display*, G02B 5/30 (349/117) (May 2000).

    An optical compensatory sheet comprises a transparent support, an orientation layer and an optically anisotropic layer in order. The optically anisotropic layer contains an aligned and fixed discotic liquid crystal compound. The orientation layer has a function of aligning the discotic liquid crystal compound. The function of the orientation layer is activated by irradiating the layer with light from a single direction. The present specification also discloses a process for the preparation of the optical compensatory sheet and a liquid crystal display using the optical compensatory sheet.

60. **US 6,064,451 (US 6,812,985)**, Y. J. Oh, K. N. Lim, D. H. Gu, S. H. Lee, and M. C. Shin, (**LG Electronics Inc.**), *Liquid crystal display device having transparent conductive layer*, G02F 1/1343 (349/40) (May 2000).

    A liquid crystal display device includes a first substrate including a plurality of data bus lines, a plurality of gate bus lines, and a plurality of thin film transistors at cross points of the plurality gate bus lines and the plurality of gate bus lines; a second substrate including a color filter layer; at least one shielding layer for shielding an outer induced electric field; and a liquid crystal layer sandwiched between the first substrate and the second substrate.

61. **US 6,066,696**, H. S. Yu and S. H. Yu, (**Samsung Display Devices Co., Ltd**), *Optical alignment composition, alignment layer formed using the same and LCD having the alignment layer*, C08L 79/00 (525/178) (May 2000).

    An optical alignment composition including a first polymer selected from a cinnamate-series polymer and a coumarin-series polymer, and a second polymer selected from polyimide having a long-chain alkyl group at its side chains and polyimide having an alkyl group at its both ends, an alignment layer formed using the same, and a liquid crystal display having the alignment layer are provided. According to the present invention, the thermal stability of the alignment layer and pretilt angle characteristic of the liquid crystal are improved.

62. **US 6,067,140**, J. W. Woo and J. B. Choi, (**LG Electronics Inc.**), *Liquid crystal display device and method of manufacturing same*, G02F 1/13 (349/129) (May 2000).

    A liquid crystal display device having a plurality of pixels, each pixel having a plurality of domains. In boundary regions between the domains, an opaque metal layer is formed to shield light transmission through these boundary regions and also to stabilize potential applied to pixel electrodes. A polyimide or photo-sensitive alignment layer is rubbed by fabric or exposed to light to provide alignment directions.

63. **US 6,083,575**, M. Ninomiya, S. Yamamoto, T. Hikichi, K. Sagawa, N. Hiji, and T. Suzuki, (**Fuji Xerox Co., Ltd**), *Polymer dispersion type liquid crystal element and manufacturing method thereof*, C09K 19/54 (428/1.1) (July 2000).

    A polymer dispersion type liquid crystal element having a sufficient reflectance and having a layered structure in which a refractive index changes periodically inside, as well as a manufacturing method thereof. The polymer dispersion type liquid crystal element is manufactured by applying laser interference light to a polymerizable composition containing a polymerizable compound having a photo-dimerizable structure and low molecular liquid crystals to conduct polymer phase separation and then applying polarized light to manufacture a polymer dispersion type liquid crystal element in which low molecular liquid crystals are oriented. A polymer dispersion type liquid crystal element having a sufficient reflectance and having a layered structure in which a refractive index changes periodically inside, as well as a manufacturing method thereof. The polymer dispersion type liquid crystal element is manufactured by applying laser interference light to a polymerizable composition containing a polymerizable compound having a photo-dimerizable structure and low molecular liquid crystals to conduct polymer phase separation and then applying polarized light to manufacture a polymer dispersion type liquid crystal element in which low molecular liquid crystals are oriented.

64. **US 6,084,057 (US 6,552,161; US 6,451,960; US 6,043,337)**, W. M. Gibbons, P. J. Shannon, and H. Zheng, (**Elsicon Inc.**), *Polarizable amines and polyimides for optical alignment of liquid crystals*, C08G 73/10 (528/353) (July 2000).

Invention relates to amine compositions and the preparation of polyimides. The polyimides can be used for inducing alignment of a liquid crystal medium with polarized light in liquid crystal display elements.

65. **US 6,091,471 (US 6,268,897; US 6,295,111; US 6,469,763; US 7,075,607)**, J. H. Kim, K. H. Yoon, J. W. Woo, M. S. Nam, Y. J. Choi, and K. J. Kim, (**LG Electronics Inc.**), *Liquid crystal cell and a method for fabricating that*, G02F1/13 (349/124) (July 2000).

    A LC cell is manufactured by the method including the steps of: rubbing a first alignment layer coating a first substrate, such that the first alignment layer has a first pretilt angle associated therewith; exposing said second alignment layer coating a first substrate to light such that said second alignment layer has at least one second pretilt angle associated therewith; and providing a liquid crystal material between said first and second substrates. The materials for the first and second alignment layers include a polyimide and a polysiloxane based material.

66. **US 6,100,953**, K. J. Kim and Y. B. Lee, (**LG. Philips LCD Co., Ltd**), *Multi-domain liquid crystal display device with concave portion in color filter and method of manufacturing thereof*, G02F 1/1337 (349/129) (August 2000).

    The present multi-domain liquid crystal display device includes first and second substrates facing each other, a liquid crystal layer between the first and second substrates, a plurality of gate bus lines arranged in a first direction on the first substrate and a plurality of data bus lines arranged in a second direction on the first substrate to define a pixel region, a thin film transistor positioned at a crossing area of the data bus line and the gate bus line and comprising a gate electrode, a semiconductor layer, and source/drain electrodes, a pixel electrode in the pixel region, a light shielding layer on the second substrate, a color filter layer having electric field-distorting recesses on its surface on the light shielding layer, a common electrode having recesses corresponding to the electric field-distorting recesses of the color filter layer, and an alignment layer on at least one substrate between the first and second substrates.

67. **US 6,103,322 (US 6,991,834; US 7,005,165; US 6,491,988; US 6,313,348)**, W. M. Gibbons, P. J. Shannon, and H. Zheng, (**Elsicon Inc.**), *Materials for inducing alignment of liquid crystals and liquid crystal optical elements*, C08G 73/10 (428/1.25) (August 2000).

The present invention provides novel polyamic acids and polyimide optical alignment layers for inducing alignment of a liquid crystal medium. The novel compositions comprise crosslinking diamines containing a C_3–C_{20} linear or branched hydrocarbon chains containing 1 to 4 carbon–carbon double bonds. The invention further describes liquid crystal displays comprising the novel polyimide optical alignment layers.

68. **US 6,107,427 (US 6,335,409)**, R. P. Herr, F. Herzog, and A. Schuster, (**Rolic AG**), *Cross-linkable, photoactive polymer materials*, C08F 220/34 (526/321) (August 2000).

    The invention is concerned with novel cross-linkable, photoactive polymer materials with 3-aryl-acrylic acid esters and amides as well as their use as orienting layers for liquid crystals and for the production of non-structured or structured optical elements and multi-layer systems.

69. **US 6,128,058**, H. G. Walton, (**Sharp KK**), *Liquid crystal device with patterned reactive mesogen alignment layer*, G02F 1/1337 (349/129) (October 2000).

    A liquid crystal device has a substrate with first and second alignment layers thereon. The first alignment layer is unidirectionally aligned. The second alignment layer is a layer formed of a cured twisted reactive mesogen composition which is patterned by masking, curing and dissolving uncured regions of the layer so as to expose first alignment regions defined by the layer, and to leave second alignment regions in the layer. The first alignment regions have a different alignment direction from that of the second alignment regions. A liquid crystal layer is provided over the resultant alignment layer structure.

70. **US 6,139,926 (US 6,348,245)**, B. C. Auman, M. P. Zussman, B. Fiebranz, and E. Bohm, (**E. I. du Pont de Nemours and Company, Merck Patent GmbH**), *Polyimide photo alignment film from 3,3,4,4-benzophenone tetracarboxylic dianhydride and ortho-substituted aromatic diamines for liquid crystal displays*, C08G 73/00 (428/1.26) (October 2000).

    The present invention relates to polyimide photo alignment films for liquid crystal display devices comprising an aromatic tetracarboxylic dianhydride component containing 3,3′,4,4′-benzophenone tetracarboxylic dianhydride and an aromatic diamine component containing at least one diamine of the group. Further display devices and methods of their fabrication employing a photo alignment process are disclosed.

71. **US 6,144,428**, M. Schadt, A. Schuster, and H. Seiberle, (**Rolic AG**), *Optical component*, G02B 5/30 (349/113) (November 2000).

    By combination of a hybrid layer formed from an orienting layer and a layer of a cross-linked liquid crystalline monomer with a reflector, and by appropriate local structuring of the molecule orientation, an optical pattern is obtained which cannot be copied using conventional photocopiers. This pattern is useful as means for protecting documents, such as banknotes, credit cards, and the like, from being copied, and can be detected by manually or automatically operated optical reading devices.

72. **US 6,169,591 (US 6,312,875; US 6,475,705)**, S. B. Kwon, Y. S. Choi, Y. Reznikov, and O. Yaroschuk, (**LG Electronics Inc.**), *Method for manufacturing a multidomain LCD panel using mask of different photo-transmittances and/or resin photosensitive to UV and visible light*, G02F 1/1337 (449/124) (January 2001).

    A method for manufacturing a multidomain liquid crystal display panel, including the steps of forming a photoalignment layer on a substrate, positioning a mask having a plurality of regions with different photo-transmittances; and forming different alignment directions in different domains of the photoalignment layer corresponding to each of the plurality of regions by irradiating the photoalignment layer with light through the mask.

73. **US 6,184,959**, Y. Izumi, (**Sharp KK**), *Liquid crystal display device having alignment film that provides alignment upon irradiation and manufacturing method the same*, G02F 1/1337 (349/124) (February 2001).

    A liquid crystal display device of the present invention includes a liquid crystal; a pair of substrates facing each other so that the liquid crystal is sandwiched therebetween and an electric field for displaying is applied to the liquid crystal; spacers fixably provided between the pair of substrates; and an alignment film material provided on at least one of the pair of substrates so as to contact the liquid crystal, the alignment film material showing aligning property upon irradiation of light. As a result, a liquid crystal display device having an excellent displaying quality is provided in which a cell gap is uniform, no contamination and impairment of an alignment film occur, and alignment nonuniformity by rubbing is not generated, even when a screen is made larger.

74. **US 6,184,961 (US 6,466,291; US 6,853,429)**, Y. S. Ham, (**LG Electronics Inc.**), *In-plane switching mode liquid crystal display device having opposite alignment directions for two adjacent domains*, G02F 1/13 (349/141) (February 2001).

    An in-plane switching mode liquid crystal display device (LCD) is disclosed in which a first portion of one pixel region has alignment direction clockwise inclined relative to the extension direction of the data electrode, and a second portion of the pixel region has alignment direction counterclockwise inclined relative to the extension direction of the data electrode. This in-plane switching mode LCD has improved viewing angle characteristics.

75. **US 6,191,836 (US 6,417,905; US 6,462,797; US 6,721,025)**, J. W. Woo, K. J. Kim, S. B. Kwon, K. H. Yoon, Y. S. Choi, J. H. Kim, and J. H. Jung, (**LG. Philips LCD Co., Ltd**), *Method for fabricating a LCC*, G02F1/1337 (349/124) (February 2001).

    A method for fabricating a liquid crystal cell and related device includes providing an alignment layer of a light sensitive material on a substrate; and exposing the alignment layer to unpolarized or partially polarized light, to provide pretilt for the molecules of the alignment layer.

76. **US 6,194,039 (US 6,242,061; US 6,380,432; US 6,713,135)**, W. M. Gibbons, P. A. Rose, P. J. Shannon, H. Zheng, (**Elsicon, Inc.**), *Materials for inducing alignment in liquid crystals and liquid crystal displays*, C08G 69/44 (428/1.1) (February 2001).

    Invention provides novel polyimides, poly(amicacides) and poly(amide esters) thereof, and optical alignment layers derived therefrom, for inducing alignment of liquid crystals and liquid crystal displays. The novel compositions comprise side-chain polymers of the general formula:

    $$—L_1—D—L_2—C_f$$

    wherein D comprises 1 to 4 carbon–carbon double bonds, C_f is a monovalent C4 to C20 fluorocarbon radical and L_1 and L_2 are linking groups. The invention further describes liquid crystal displays and other liquid crystal devices.

77. **US 6,201,087**, R. P. Herr, F. Herzog, and A. Schuster, (**Rolic AG**), *Coumarin and quinolinone derivatives for the production of orienting layers for liquid crystals*, C08F 220/58 (526/268) (March 2001).

Invention provides a novel photochemically dimerizable coumarin or quinoline derivative.

78. **US 6,203,866 (US 6,395,352)**, A. Mochizuki, T. Yoshihara, H. Shiroto, T. Makino, and Y. Kiyota, (**Fujisu Ltd**), *Ferroelectric liquid crystal display element and manufacturing method thereof*, G02F 1/141 (428/1.1) (March 2001).

    When a photopolymerization compound (molecular weight: not less than 300) having at least two ultraviolet light sensitive portions in its molecular structure is mixed with ferroelectric liquid crystal and an exposure to UV light is performed, a continuous structure is formed over almost whole length between two glass substrates, and a continuous twisted structure is formed between the glass substrates in the state of liquid crystal molecular orientation. Before the exposure to ultraviolet light, the photopolymerization compound does not exhibit satisfactory compatibility with the ferroelectric liquid crystal to be mixed, but exhibits phase separation in the mixed state. The ferroelectric liquid crystal has a bookshelf layer structure.

79. **US 6,215,539 (US 5,838,407; US 6,300,991; US 6,608,661)**, M. Schadt, A. Schuster, and H. Seiberle (**Rolic AG**), *Photo-oriented polymer network material having desired azimuthal orientation and tilt angle and method for its production*, G02F1/1337 (349/124) (April 2001).

    Orientation layers of monomeric or polymeric liquid crystal layers having any desired azimuthal orientation direction and a tilt angle are produced by irradiating at an angle differing from the normal to the surface photopolymers orienting parallel to linearly polarized light.

80. **US 6,218,501**, H. J. Choi and J. Y. Kim, (**Cheil Industries Inc.; Samsung Electronics Co., Ltd**), *Polymaleimide and polyimide photoalignment materials for LCD, and polyimide photoalignment materials for LC display*, C08G 73/06 (528/170) (April 2001).

    Invention relates to a cinnamatic photo-polymerization type homopolymeric or copolymeric alignment material, in which polymaleimide is singly used as the main chain, or is combined with styrene, hydroxystyrene or axrylonitrile to form a copolymer so as to be used as the main chain, or polyimide is used as the main chain.

81. **US 6,222,601**, Y. J. Choi and J. H. Kim, (**LG Electronics Inc.**), *Method of manufacturing a liquid crystal cell*, G02F 1/1337 (349/129) (April 2001).

A method of manufacturing a liquid crystal cell in accordance with the present invention comprises the steps of forming a first alignment layer and a second alignment layer on a first substrate and a second substrate, forming a pretilt having an alignment, pretilt angle direction, and pretilt angle of the first alignment layer, forming a liquid crystal layer between the first substrate and second substrate after sealing them.

82. **US 6,224,788**, T. Ogawa, S. Nakata, Y. Makita, M. Kimura, Y. Takeuchi, Y. Matsuki, and S. I. Kimura, (**JSP Co.**), *Liquid crystal aligning agent and process for producing liquid crystal alignment film using the same*, G02F 1/1337 (252/299.4) (May 2001).

    A liquid crystal aligning agent composed of a polymer containing a moiety having two aromatic nuclei and an ethylene carbonyl group therebetween. The liquid crystal aligning agent is converted to a liquid crystal alignment film by irradiating a thin film of the agent with a linearly polarized ray.

83. **US 6,226,066 (US 6,414,734; US 6,433,850; US 6,633,355; US 6,879,363; US 7,145,618)**, Y. Reznikov, O. Yaroschuk, J. W. Woo, Y. J. Choi, K. H. Yoon, M. S. Nam, J. H. Kim, and S. B. Kwon, (**LG. Philips LCD Co., Ltd**), *Method for controlling pretilt angle direction in a liquid crystal cell*, G02F1/1337 (349/124) (May 2001).

    A method is disclosed for controlling a pretilt angle direction for a liquid crystal cell comprising the steps of first setting the magnitude of pretilt angle and a plurality of pretilt angle directions in an alignment layer. This first step is achieved by irradiating linearly the alignment layer with polarized or unpolarized UV light. One of the plurality of pretilt angle directions is then selected by exposing the alignment layer to UV light a second time.

84. **US 6,243,151 (US 6,445,431; US 6,501,520)**, T. Nose, S. Kaneko, and M. Suzuki, (**Nippon Electric Co.**), *Liquid crystal display with polarization layer interior to substrates*, G02F 1/1335 (349/70) (June 2001).

    There is provided a liquid crystal display including (a) a backlight source having a dominant emission peak at 380–420 nm, (b) a first polarization layer for selecting a light directed in a predetermined direction among lights emitted from the backlight source, (c) a second polarization layer for receiving a light selected by the first polarization layer, (d) first and second transparent substrates, (e) first and second transparent electrodes, (f) a liquid crystal layer, and (g) a fluorescent material layer receiving lights from the backlight source and emitting a light therefrom. The second polarization layer is located intermediate between the first and second transparent substrates. The backlight source, the first polarization layer, the first transparent substrate, the first transparent electrode, the liquid crystal layer, the second transparent electrode, the second polarization layer, the fluorescent material layer, and the second transparent substrate are preferably deposited from bottom to top in this order. The above-mentioned liquid crystal display makes it possible to reduce the number of transparent substrates by one relative to a conventional liquid crystal display, which ensures higher brightness and no parallax.

85. **US 6,259,502 (US 6,384,888; US 6,741,312)**, H. Komatsu, (**LG Electronics Inc.**), *In-plane switching mode liquid crystal display device having a common electrode on the passivation layer*, G02F 1/1343 (349/141) (July 2001).

    An in-plane switching mode liquid crystal display device comprises a substrate, a pixel region, a common bus line, a thin film transistor, a data electrode, a passivation layer over the data electrode and the thin film transistor, and a common electrode. The pixel region lies on the substrate. The common bus line is aligned in the pixel region. The thin film transistor is coupled to the pixel region and the pixel region comprises a gate electrode and a gate insulator having a portion overlying the gate electrode. The data electrode lies over the gate insulator and has a portion overlying the common bus line to form a first storage capacitor. The passivation layer overlies the data electrode and the thin film transistor. The common electrode overlies the passivation layer and has a portion overlying the data electrode to form a second storage capacitor.

86. **US 6,277,502**, R. Buchecker, G. Marck, and H. Seiberle, (**Rolic AG**), *Photocrosslinkable silane derivatives*, C07F 7/12 (428/689) (August 2001).

    The invention relates to novel crosslinkable, photoactive silane derivatives of the formula I,

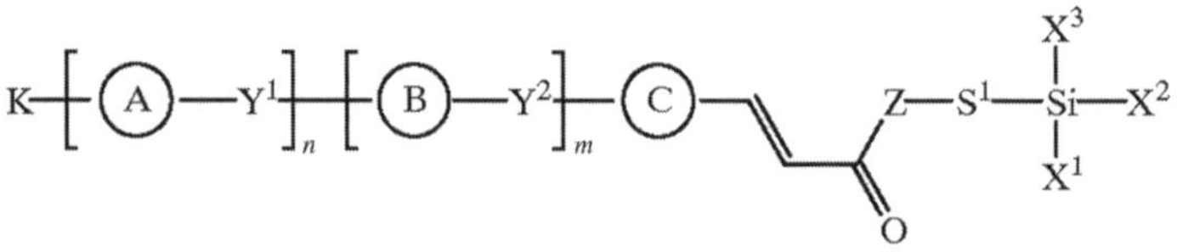

K—[(A) —Y¹]ₙ—[(B) —Y²]ₘ— (C) —CH=CH—C(=O)—Z—S¹—Si(X¹)(X²)(X³)

mixtures of silane derivatives of the formula I and mixtures of silane derivatives of the formula I with uncrosslinkable silane derivatives as usually used for silanizing inorganic, oxide-containing surfaces. The invention furthermore relates to the use of silane derivatives of the formula I and of mixtures which contain at least one silane derivative of the formula I as orientation layers for liquid crystals and for the production of unstructured or structured optical elements and multilayer systems.

87. **US 6,284,197 (US 6,852,285; US 6,858,423)**, N. L. Abbott, J. J. Skaife, V. K. Gupta, T. B. Dubrovsky, and R. Shah, (**The Regents of the University of California**), *Optical amplification of molecular interactions using liquid crystals*, G01N 21/75 (422/82.5) (September 2001).

    Interactions between molecules which are components of self-assembled monolayers and other molecules can be amplified and transduced into an optical signal through the use of a mesogenic layer. The invention provides a device and methods for detecting analytes. The device comprises a substrate onto which a self-assembled monolayer is attached and a mesogenic layer which is anchored by the self-assembled monolayer. The mesogenic layer undergoes a change in conformation in response to the molecular interaction.

88. **US 6,292,244 (US 6,300,993; US 6,377,326)**, C. L. Kuo, H. D. Liu, and C. K. Wei, (**Industrial Technology Research Institute**), *Manufacturing method of forming two-domain liquid crystal display by exposing a part of the orientation layer twice*, G02F 1/13 (349/129) (September 2001).

    The present invention is provided a two-domain vertical aligned LCD with negative compensator. The liquid crystal molecules in each of the liquid crystal domains are orientated nearly perpendicular to surfaces of the transparent substrates with a little pre-tilted angle to the normal of said substrates when an electric field is not applied, the tilt angle projected on the azimuthal of substrate between the orientation of liquid crystal molecules in two domains are not equal to 180 degrees, and the orientation of said liquid crystal molecules in two domains are oriented opposed direction. The structure that described above is obtained by UV exposure process.

89. **US 6,292,296 (US 6,639,720; US 7,016,112; US 7,016,113; US 7,061,679)**, J. B. Choi, B. D. Song, and K. H. Yoon, (**LG. Philips LCD Co., Ltd**), *Large scale polarizer and polarizer system employing it*, G.02B 5/30 (359/487) (September 2001).

A large scale polarizer comprises one or more quartz substrate parts formed as a rectangle, a triangle, or a parallelogram, and a polarizer holder supporting the quartz substrate part. The polarizer holder may be in a lattice structure holding a plurality of quartz substrate parts. A polarizer system employing the large scale polarizer comprises a lens making an incident light to a parallel light, the large scale polarizer, and a moving control part coupled to and moving the large scale polarizer.

90. **US 6,297,866 (US 6,798,482)**, S. M. Seo, Y. S. Ham, J. J. Park, H. H. Shin, and Y. S. Ahn, (**LG. Philips LCD Co., Ltd**), *In-plane switching mode liquid crystal display device*, G02F 1/13 (349/141) (October 2001).

    An in-plane switching mode liquid crystal display device includes first and second opposed substrates having inner surfaces in which a liquid crystal layer formed therebetween, a data bus line and a gate bus line arranged perpendicularly and/or horizontally in a matrix on the first substrate thereby defining a unit pixel region, and a pair of data electrode and common electrode applying a plane electric field in the liquid crystal layer, the electrodes being inclined with respect to the data bus line and parallel to each other.

91. **US 6,307,609 (US 6,407,894; US 6,061,138)**, W. M. Gibbons and B. P. McGinnis, (**Elsicon Inc.**), *Polarized light exposure systems for aligning liquid crystals*, G02F 1/13, (349/124) (October 2001).

    This invention relates to an optical exposure system that is useful for exposing optical alignment layers with light in order to align liquid crystal. The exposure system comprises at least one source of optical radiation, means for partially collimating the optical radiation, means for partially polarizing the optical radiation, means for adjusting the degree of polarization, and means for transporting a substrate and radiation relative to one another. Other embodiments include means for partially filtering the optical radiation and means for some portion of said optical radiation to be incident at an oblique angle relative to the substrate. Another embodiment is a novel optical module and processes for aligning liquid crystals using the optical exposure systems.

92. **US 6,312,769**, H. Hiraoka, Y. Takeuchi, S.-I. Kimura, Y. Matsuki, T. Ogawa, and M. Kimura, (**JSR Corporation**), *Liquid crystal alignment layer, production method for the same, and liquid crystal display device comprising the same*, G02F 1/13 (428/1.1) (November 2001).

A method of producing a liquid crystal alignment layer comprising exposing an organic film to polarized pulse laser beam to align molecules in a surface portion of the organic film. There is provided a liquid crystal alignment layer composed of a polyamide film having aligned molecules in a surface portion.

93. **US 6,335,776**, K. J. Kim and J. J. Yoo, (**LG. Philips LCD Co., Ltd**), *Multi-domain liquid crystal display device having an auxiliary electrode formed on the same layer as the pixel electrode*, G02F 1/13 (349/129) (January 2002).

    A multi-domain liquid crystal display device includes first and second substrates facing each other and a liquid crystal layer between the first and second substrates. A plurality of gate bus lines are arranged in a first direction on the first substrate and a plurality of data bus lines are arranged in a second direction on the first substrate to define a pixel region. A common electrode is formed on the second substrate and a pixel electrode is electrically charged through the data bus line to drive the liquid crystal layer with the common electrode. An auxiliary electrode is formed on a same layer whereon the pixel electrode is formed. A passivation is formed below the pixel and auxiliary electrodes to insulate them from other electrodes and bus lines. An alignment layer is on at least one substrate between the first and second substrates.

94. **US 6,368,681**, K. Ogawa, (**Matsushita Electric Industrial Co., Ltd**), *Liquid crystal alignment film, method of manufacturing the film, liquid crystal display using the film and method, and method of manufacturing the liquid crystal display*, G02F 1/1337 (428/1.23) (April 2002).

    A positive resist mainly composed of a novolak resin and comprising a naphthoquinone diazido-based photosensitizer as an energy beam sensitive resin (e.g., a photosensitive resin) is applied in a thickness of 0.1 to 0.2 μm to a surface of a glass substrate provided with transparent electrodes and dried so as to form a photosensitive film. Next, using a mask, the film is exposed to ultraviolet rays (365 nm). Then, moisture in the air reacts with the resist in an exposed portion 2′, thereby generating —COOH groups, with which $CH_3(CH_2)_{18}SiC_{13}$ is allowed to react so as to cause a dehydrochlorination reaction, thereby forming a monomolecular chemisorption film comprising carbon chains. This film is used as an alignment film. Thus, the present invention provides a method for producing a uniform and thin alignment film for use in a liquid crystal display panel with a high efficiency without performing a rubbing treatment, and a method for producing a display panel using the same.

95. **US 6,383,579 (US 6,582,784; US 7,018,687)**, S. H. Park and Y. S. Choi, (**LG. Philips LCD Co., Ltd**), *Liquid crystal display device*, G02F1/1337 (428/1.26) (May 2002).

    The liquid crystal display device of the present invention comprises first and second substrates, a first alignment layer on the first substrate, wherein the first alignment layer includes the photo-sensitive constituent includes a material selected from the group consisting of cinnamoyl derivatives or polyethyleneimine, and a liquid crystal layer between the first and second substrates.

96. **US 6,399,165 (US 6,572,939)**, S. B. Kwon, K. J. Kim, Y. S. Choi, I. Gerus, A. Dyadysha, and Y. Reznikov, (**LG. Philips LCD Co., Ltd**), *Liquid crystal display device*, G02F1/1337 (428/1.2) (June 2002).

    A liquid crystal display device comprises first and second substrates, an alignment layer including a pyranose polymer or a furanose polymer on at least one of the first and second substrates, and a liquid crystal layer between the first and second substrates. The liquid crystal display device is characterized by excellent thermostability, superior anchoring energy and uniform alignment of the liquid crystal achieved in a reduced treatment time without creating any flowing effect in the liquid crystal.

97. **US 6,445,435 (US 6,281,959; US 6,396,558; US 6,628,362; US 6,697,140; US 6,822,717)**, S. M. Seo, H. H. Shin, Y. J. Oh, H. C. Lee, and C. Y. Kim, (**LG. Philips LCD Co., Ltd**), *In-plane switching mode liquid crystal display device having common electrode on passivation layer*, G02F 1/1362 (349/141) (September 2002).

    An in-plane switching mode liquid crystal display device comprises first and second substrates, a plurality of gate and data bus lines defining pixel regions and arranged on the first substrate, a plurality of data electrodes on same plane of the data bus lines these some parts are overlapped with adjacent gate bus line, a passivation layer on the data electrodes, a plurality of common electrodes on the passivation layer these some parts are overlapped with adjacent data electrodes, and a liquid crystal layer between the first and second substrates.

98. **US 6,449,025 (US 6665035)**, Y. B. Lee, (**LG. Philips LCD Co., Ltd**), *Multi-domain liquid crystal display device having field affecting electrode*, G02F 1/13 (349/129) (September 2002).

    The present multi-domain liquid crystal display device includes first and second substrates facing each other, a liquid crystal layer between the first and second substrates, a plurality of gate bus lines arranged in a first direction on the first substrate and a plurality of data bus lines arranged in a second direction on the first substrate to define a pixel region, a thin film transistor positioned at a crossing area of the data bus line and the gate bus line and comprising a gate electrode, a semiconductor layer, and source/drain electrodes, a pixel electrode in the pixel region, a subsidiary electrode on the second substrate, a color filter layer on the subsidiary electrode, a common electrode on the color filter layer, and an alignment layer on at least one substrate between the first and second substrates.

99. **US 6,459,463**, Y. B. Kim, J. M. Moon, S. J. Bae, and J. H. Jun, (**LG. Philips LCD Co., Ltd**), *Reflective liquid crystal display having a bent shape and method of manufacturing thereof*, G02F 1/1362 (349/113) (October 2002).

    A reflective-type liquid crystal display device includes first and second substrates, a liquid crystal layer between the first and second substrates, thin film transistors on each cross of gate and data bus lines, a photoresist layer on the gate and data bus lines, and thin film transistors, reflection electrodes on the photoresist layer and electrically coupled to the thin film transistors, and light shield layers on the gate and data bus lines.

100. **US 6,461,694**, H. Nishikawa and K. Kawata, (**Fuji Photo Film Co., Ltd**), *Optical compensatory sheet comprising transparent support, orientation layer and optically anisotropic layer*, C09K 19/32 (428/1.3) (October 2002).

     An optical compensatory sheet comprises a transparent support, an orientation layer and an optically anisotropic layer in this order. The optically anisotropic layer is formed from discotic liquid crystal molecules. The discotic liquid crystal molecule comprises a triphenylene nucleus and six cinnamoyl side chains. The optically anisotropic layer has a high refractive anisotropy in the range of 0.065 to 0.16.

101. **US 6,466,286**, H. S. Seo, (**LG. Philips LCD Co., Ltd**), *Reflecting substrate for a liquid crystal display apparatus including an aluminum neodymium electrode and manufacturing method thereof*, G02F 1/1335 (349/113) (October 2002).

A reflectance-type liquid crystal display apparatus includes a gate electrode, a gate bus line, and a gate pad electrode which are formed on a substrate and made of AlNd. An insulating layer is disposed on the gate electrode, the gate bus line and the gate pad electrode. An AlNd reflecting electrode is disposed on the insulating layer.

102. **US 6,466,288**, B. K. Rho, (**LG LCD Inc.**), *Multi-domain liquid crystal display device*, G02F 1/139 (349/141) (October 2002).

 A multi-domain vertical alignment liquid crystal display device includes upper and lower substrates having upper and lower alignment layers, a pixel electrode on the lower substrate, and a liquid crystal layer between the upper and lower alignment layers. The upper alignment layer is alignment-treated to define a specified alignment direction. The pixel electrode is arranged to define a desired pattern of slits or floating electrodes. The liquid crystal layer includes liquid crystal molecules having negative dielectric anisotropy.

103. **US 6,473,142 (US 6,356,335; US 6,525,794; US 6,825,906)**, K. J. Kim, J. J. Yoo, and S. J. Bae, (**LG. Philips LCD Co., Ltd**), *Multi-domain liquid crystal display device*, G02F 1/1333 (349/84) (October 2002).

 A multi-domain liquid crystal display device comprises first and second substrates facing each other and a liquid crystal layer between the first and second substrates. A plurality of gate bus lines are arranged in a first direction on the first substrate and a plurality of data bus lines are arranged in a second direction on the first substrate to define a pixel region. A pixel electrode electrically is charged through the data bus line in the pixel region, a color filter layer is formed on the second substrate, and a common electrode is formed on the color filter layer. Dielectric frames are formed in the pixel region, and an alignment layer on at least one substrate between the first and second substrates.

104. **US 6,479,218 (US 6,787,292)**, Y. S. Choi, (**LG. Philips LCD Co., Ltd**), *Method for manufacturing multi-domain liquid crystal cell*, G02F1/1337 (430/321) (November 2002).

 A method of manufacturing a multi-domain liquid crystal display device having a pixel includes the steps of forming an alignment film on at least one of a first and second substrate; covering the alignment film with a mask, there being included a first surface having a plurality of light-transmitting portions and light-shielding portions and a second surface having light-shielding portions corresponding to the light-transmitting portions; radiating light from an upper portion of the mask; and assembling the first and second substrates.

105. **US 6,496,239**, H. Seiberle, (**Rolic AG**), *Optical component for producing linearly polarized light*, G02F 1/1335 (349/98) (December 2002).

A polarization mask for producing light of locally different polarization from unpolarized or uniformly polarized light. On the light output side, discrete areas with different polarization directions are present, which are either static or switchable. The mask is useful for, among other things, the transfer of polarization pattern onto a PPN layer.

106. **US 6,504,591 (US 7,088,412)**, K. Kondo, Y. Tomioka, T. Miwa, and M. Yoneya, (**Hitachi, Ltd**), *Liquid crystal display apparatus*, G02F 1/13 (349/123) (January 2003).

In one pixel, a stepped difference portion having an inclination degree a/b is formed on an alignment control layer, the alignment control layer is made of a material in which a liquid crystal alignment capability is given according to a polarization light radiation, and according to the polarization light radiation the liquid crystal alignment capability is added in the vicinity of the stepped difference portion. A thick and transparent organic high molecular layer is interposed between the alignment control layer and a substrate and a close adhesion characteristic and a transparency characteristic can be made compatible therewith. Accordingly, a problem in which the manufacturing margin in an alignment processing is narrow can be solved, and generation of a display failure according to a fluctuation in the initial alignment direction can be reduced. In this way, a large size liquid crystal display apparatus having a high quality and in which the contrast ratio is heightened can be provided.

107. **US 6,512,564 (US 6,781,656; US 7,133,099)**, H. Yoshida, T. Seino, and Y. Tasaka, (**Fujitsu Limited**), *Alignment treatment of liquid crystal display device*, G02F 1/13 (349/124) (January 2003).

The liquid crystal display device comprising a pair of substrates with alignment layers formed thereon, and a liquid crystal filled between the substrates. Each pixel has pixel display portions CA, CB and non-display portions DA, EA, DB, EB. The pixel display portions are treated for realizing alignment in a different manner from the non-display portions and the alignment of the pixel display portions is controlled by the alignment of the non-display portions. Moreover, the alignment treatment is executed by the irradiation with ultraviolet rays in an inclined direction.

108. **US 6,549,258**, H. H. Shin, J. J. Park, and W. S. Kim, (**LG. Philips LCD Co., Ltd**), *Hybrid switching mode liquid crystal display device*, G02F 1/13 (349/141) (April 2003).

    A hybrid switching mode liquid crystal display device includes first and second substrates and a liquid crystal layer between the first and second substrates. A plurality of first electrodes are formed on the first substrate and applies an in-plane electric field. A plurality of second electrodes are formed on the second substrate and applies perpendicular and inclined electric fields with the first electrodes.

109. **US 6,569,972**, H. J. Choi, J. L. Kim, E. K. Lee, and J. Y. Kim, (**Samsung Electronics Co.**), *Photoalignment materials for liquid crystal alignment film*, C07D 207/40 (526/262) (May 2003).

    A photoalignment material useful in liquid crystal alignment films comprises a maleimide-based repeating unit and at least one additional repeating unit, or a maleimide-based repeating unit and at least two additional repeating units. The photoalignment materials have freely-controllable pretilt angles, and they provide a display quality equivalent or superior to alignment materials made using the conventional rubbing process.

110. **US 6,582,775**, J. A. Payne and J. W. Hoff, (**Eastman Kodak Co.**), *Process for making an optical compensator film comprising photoaligned orientation layer*, C09K 19/02 (427/508) (June 2003).

    A process for making an optical compensator comprising the steps of coating a photo-alignable resin in a solvent onto a substrate; drying the resin-containing coating to form an orientable layer; heat treating the orientable layer before and/or after orientation, orienting the orientable layer in a predetermined direction; coating an anisotropic liquid crystal compound in a solvent carrier onto the orientation layer; drying the anisotropic layer. In a preferred embodiment, these steps are repeated so that the optical axis of a first anisotropic layer is positioned orthogonally relative to the respective optical axis of a second anisotropic layer about an axis perpendicular to the plane of the substrate.

111. **US 6,582,776**, W. C. Yip, E. K. Prudnikova, H. S. Rwok, V. G. Chigrinov, V. M. Kozenkov, H. Takada, and M. Fukuda, (**Hong Kong University of

Science and Technology; Dainippon Ink and Chemicals, Inc.), *Method of manufacturing photoalignment layer*, C08J7/18, G02F 1/13 (427/514) (June 2003).

A photoalignment layer having excellent long-term stability to light and heat is manufactured by coating a material for the photoalignment layer, which contains a dichroic dye having two or more polymerizable groups per molecule, on a substrate, and exposing the coating layer to polarized light, thereby imparting a photo-alignmenr function, and polymerizing the polymerizable groups by heating or light exposure.

112. **US 6,583,835**, H. Yoshida, Y. Tasaka, T. Sasabayashi, Y. Nakanishi, and K. Okamoto, (**Fujitsu Display Technologies Co.**), *Alignment films in a liquid crystal display device and a method of manufacturing the same*, G02F 1/13 (349/124) (June 2003).

A liquid crystal display device including a pair of substrates in a spaced relationship with one another. A pair of alignment films are provided, one alignment film being formed on each substrate such that the alignment films face one another. A liquid crystal layer, including plural liquid crystals, is inserted between the pair of alignment films, wherein the alignment films impart a given pre-tilt angle to the liquid crystals. The alignment films are composed of a material containing at least two types of polymers having a prescribed initial alignment and different alignment variation rates in response to ultra-violet ray irradiation. The pre-tilt angle being adjusted, without rubbing the alignment films, through UV exposure of the alignment films.

113. **US 6,590,627 (US 6,958,799)**, Y. Tomioka, Y. Umeda, and K. Kondo, (**Hitachi Ltd**), *Liquid crystal display*, G02F 1/13 (349/139) (July 2003).

The present invention provides a liquid crystal display of high image quality which can be driven at a low voltage and shows a lowered inhomogeneity of display caused by an after image phenomenon. The liquid crystal display comprises one pair of substrates which are a first substrate and a second substrate, at least one of which is transparent, a liquid crystal layer and a color filter layer provided between the two substrates, a plurality of thin film transistors provided on the first substrate and connected to an image signal wiring and a scanning signal wiring, a common electrode giving a standard potential, and a pixel electrode connected to the thin film transistors and placed opposite to the common electrode in a pixel region, wherein the common electrode and pixel electrode are placed in different layers from each other through an interlaminar insulating film in the form of a layer comprising at least two layers comprising the color filter layer.

114. **US 6,593,989 (US 6,661,489)**, J. Ha Kim and J. Hy Kim, (**LG Electronics Inc.**), *Domain-divided twisted nematic liquid crystal cell and method of fabricating thereof*, G02F 1/13 (349/129) (July 2003).

The method of the present invention comprises the steps of providing first and second substrates, forming a photoalignment layer in each domain of the first substrate having two or more domains, wherein thicknesses of the photoalignment layers corresponding to the domains are different from each other. A pretilt angle in each domain is formed by light irradiation on the photoalignment layer, wherein the pretilt angles corresponding to the domains are different from each other due to the different thicknesses of the photoalignment layer in different domains. The first and second substrates are positioned to face each other, and liquid crystal is injected between the first and second substrates.

115. **US 6,597,422**, J. Funfschilling, M. Stadler, and M. Schadt, (**Rolic AG**), *Orientational layer for a liquid crystal material*, G02F 1/13 (349/123) (July 2003).

A liquid crystal device comprising ferroelectric liquid crystal material aligned by a liquid crystal polymer (LCP) network layer under 20 nm thick, which itself was aligned by a photo-oriented linearly photopolymerized (LPP) layer under 20 nm thick, exhibits low voltage drop over the aligning layer and has a remarkable contrast ratio.

116. **US 6,610,462**, L. C. Chien and O. Yaroshchuk, (**Rolic AG**), *Liquid crystal alignment using photo-crosslinkable low molecular weight materials*, G02F1/1337 (430/321) (August 2003).

A method is provided to align liquid crystal in liquid crystal displays by using photo-crosslinkable low molecular weight materials. The material contains a polymerizable group and a photosensitive group. The material is dissolved in a solvent and is disposed onto a substrate and the solvent is removed. The resulting film is then cured with UV light to cross-link and induce anisotropy in the material. Also, a liquid crystal display that contains an alignment layer that is the photo-reaction product of a photo-crosslinkable material that contains at least one polymerizable group and at least one photosensitive group.

117. **US 6,620,920 (US 6,046,290)**, H. Berneth, U. Claussen, S. Kostromine, R. Neigl, J. Rubner, and R. Ruhmann, (**Bayer Aktiengesellschaft**), *Monomers for photoaddressable side group polymers of high sensitivity*, C07C 311/37 (534/574) (September 2003).

    The invention provides an optical method which permits, with the aid of 6 simple measurements, a conclusion regarding the suitability of antennas (groups which can absorb electromagnetic radiation) for incorporation into photoaddressable polymers.

118. **US 6,627,269 (US 6,797,096; US 7,220,467)**, M. S. Nam, (**LG. Philips LCD Co., Ltd**), *Photoalignment material and liquid crystal display device and its manufacturing method using the same*, G02F1/13 (428/1.2) (September 2003).

    A photoalignment material, a liquid crystal display device using the photoalignment material, and a manufacturing method. The photoalignment material is a polymer having a photo-reactive ethenyl group on a main chain. When used as a photoalignment layer, the photoalignment material enables improved alignment stability against external shocks, light, and heat. The liquid crystal display device includes a first substrate, a second substrate, a liquid crystal layer formed between the first and second substrates, and a photoalignment layer formed at least on the first substrate, with the photoalignment layer formed from a photoalignment material having an ethenyl group at a main chain.

119. **US 6,630,289**, H. S. Kwok, W. C. Yip, V. G. Chigrinov, and V. Kozenkov, (**Hong Kong University of Science and Technology**), *Photo-patterned light polarizing films*, G03F 7/00 (430/326) (October 2003).

    This invention relates to methods for preparing photo-patterned mono- or polychromatic, polarizing films. The polarizer can be pixelated into a number of small regions wherein some of the regions have one orientation of the principal neutral or color absorbing axis; and some other of the said regions have another orientation of the principal neutral or color absorbing axis. The axis orientation is determined by the polarization vector of actinic radiation and the multi-axes orientation is possible by a separated masked exposure. This polarizer can be placed on the interior substrate surface of the LCD cell.

120. **US 6,646,703**, H. Seiberle and M. Schadt, (**Rolic AG**), *Method of making an element of liquid crystal polymer using alignment induced by unpolarized or circularly polarized radiation*, G02B 5/30 (349/124) (November 2003).

 A method of imparting a property to a material, the property being that cross-linkable monomeric or pre- polymeric liquid crystal molecules which may be placed on the layer would adopt preferred alignment. The method comprises exposing the material to unpolarized or circularly polarized radiation from an oblique direction, wherein the material is, for example, cross-linked by the irradiation, allowing monomeric or pre-polymeric liquid crystal molecules applied to or mixed with the exposed material to adopt the preferred alignment and while aligned cross-linking them. The invention also relates to LCD elements incorporating a preferred alignment.

121. **US 6,654,090**, K. J. Kim, Y. B. Lee, J. J. Yoo, D. H. Kwon, J. B. Choi, and Y. I. Park, (**LG. Philips LCD Co., Ltd**), *Multi-domain liquid crystal display device and method of manufacturing thereof*, G02F 1/13 (349/129) (November 2003).

 The present multi-domain liquid crystal display device includes first and second substrates facing each other; a liquid crystal layer between the first and second substrates; a plurality of gate bus lines arranged in a first direction on the first substrate and a plurality of data bus lines arranged in a second direction on the first substrate to define a pixel region; a thin film transistor positioned at a crossing area of the data bus line and the gate bus line, the thin film transistor comprising a gate electrode, a semiconductor layer, and source/drain electrodes; a pixel electrode on the first substrate, the pixel electrode having at least one window inducing electric field therein; a color filter layer on the second substrate, the color filter layer having at least one window distorting electric field therein; a common electrode on the color filter layer; and an alignment layer on at least one substrate between the first and second substrates.

122. **US 6,661,489 (US 6,593,989)**, J. Ha. Kim and J. Hy. Kim, (**LG Electronics Inc.**), *Domain-divided twisted nematic liquid crystal cell and method of fabricating thereof having a differentially removed photosensitive layer*, G02F1/1337 (349/124) (December 2003).

 A domain-divided twisted nematic crystal cell and method of fabricating thereof. The method of the present invention comprises the steps of providing first and second substrates, forming a photoalignment layer in each domain of the first substrate having two or more domains, wherein thicknesses of the photoalignment layers corresponding to the domains are different from each other. A pretilt angle in each domain is formed by light irradiation on the photoalignment layer, wherein the pretilt angles corresponding to the domains are different from each other due to the different thicknesses of the photoalignment layer in different domains. The first and second substrates are positioned to face each other, and liquid crystal is injected between the first and second substrates.

123. **US 6,671,021 (US 6,882,391)**, M. H. Lee, (**LG. Philips LCD Co., Ltd**), *HTN mode liquid crystal display device*, G02F 1/1337 (349/129) (December 2003).

 A homeotropic twisted nematic (HTN) mode LCD device is disclosed, which obtains a multi-domain, a wide viewing angle, and a rapid response time by applying photoalignment to the HTN mode. The HTN mode LCD device includes first and second substrates, a first alignment film on the first substrate, the first alignment film having at least two domains and having a different alignment direction in the domain, a second alignment film on the second substrate, having the same alignment direction in the domain, and a liquid crystal layer between the first and second substrates.

124. **US 6,686,980**, M. Ichihashi, (**Fuji Photo Film Co Ltd**), *Anisotropic film containing a copolymer including a monomer containing a dichroic dye and liquid crystal display containing the anisotropic film*, C09K 19/60 (349/96) (February 2004).

 An anisotropic film is described, which is a hard film with maintaining the orientation of a polymerizable dichroic dye and forming a copolymer with other polymerizable monomer.

125. **US 6,731,362**, K. J. Park, H. T. Kim, J. W. Lee, and S. J. Sung, (**Korea Advanced Institute of Science and Technology**), *Polymer blend for preparing liquid crystal alignment layer*, C08L 79/00 (349/123) (May 2004).

 The present invention relates to a polymer blend of cinnamate polymer and polyimide polymer, which are photo-reactive polymers, for preparing liquid crystal alignment layer having a high pretilt angle in photoalignment, a process for preparing liquid crystal alignment layer by employing the said blend, a liquid crystal alignment layer prepared by the process, and a crystal cell prepared by employing the liquid crystal alignment layer. The polymer blend for preparing liquid crystal alignment layer of the invention comprises 10 to 90% (w/w) of cinnamate polymer and 10 to 90% (w/w) of polyimide polymer. Since the liquid crystal alignment layer prepared by employing the polymer blend of the invention has an excellent alignment property and thermal stability, which makes possible its wide application in the development of liquid crystal displays.

126. **US 6,733,958 (US 6,784,231)**, M. Fukuda, H. Hayakawa, and H. Takada, (**Dainippon Ink and Chemicals, Inc.**), *Material for photoalignment layer, photoalignment layer and method of manufacturing the same*, C07D 207/452 (430/321) (May 2004).

    The present invention provides a photoalignment layer for a liquid crystal display device, which has good liquid crystal display device characteristics such as a good voltage holding ratio and also has good alignment stability and sufficient resistance to light and heat. The photoalignment layer is manufactured by coating a polymerizable monomer having at least one photoalignment moiety, which carries out a photoalignment function by the photo dimerization reaction, and at least two polymerizable maleimide groups per molecule on a substrate, and exposing the coating layer to light to cause the photo dimerization reaction of the structural unit and the photopolymerization reaction of the polymerizable maleimide group, thereby to form a crosslinked polymeric layer and to enable the polymeric layer to carry out the photoalignment function.

127. **US 6,737,303**, H. L. Cheng, W. Y. Chou, C. Y. Sheu, Y. W. Wang, J. C. Ho, and C. C. Liao, (**Industrial Technology Research Institute (TW)**), *Process for forming organic semiconducting layer having molecular alignment*, H01L 51/40, (438/150) (May 2004).

    A process for forming an organic semiconducting layer having molecular alignment. First, a photoalignment organic layer is formed on a substrate or a dielectric layer. Next, the photoalignment organic layer is irradiated by polarized light through a layer having molecular alignment. Finally, an organic semiconducting layer is formed on the orientation layer, such that the organic semiconducting layer aligns according to the alignment of the orientation layer to exhibit molecular alignment. The present invention can form an organic semiconducting layer with different molecular alignments in different regions over the same substrate by means of polarized light exposure through a mask.

128. **US 6,749,895**, K. J. Park, S. J. Sung, and J. W. Lee, (**Korea Advanced Institute of Science and Technology**), *Polymer for preparing liquid crystal alignment layer*, C08G 73/00 (427/335) (June 2004).

The present invention provides a polymer in which coumarin, a photo-reactive molecule, is grafted onto a polyimide for preparing liquid crystal alignment layer which has a superior alignment property and an excellent thermal stability in photoalignment, a process for preparing the said grafted polymer, a process for preparing liquid crystal alignment layer by employing the said grafted polymer, and a liquid crystal alignment layer prepared by the process. The polymer of the invention is prepared by mixing a coumarin compound with a polyimide, dissolving the mixture in an organic solvent, adding a catalyst, and stirring under an environment of N_2 gas. The polymer of the invention is superior in terms of the thermal stability, which makes possible its universal application for the development of a novel liquid crystal display (LCD).

129. **US 6,751,003 (US 6,914,708)**, X. D. Mi, (**Eastman Kodak Co.**), *Apparatus and method for selectively exposing photosensitive materials using a reflective light modulator*, G02F 1/13 (359/247) (June 2004).

An exposure system for fabricating optical film by exposing a pattern such as for photoalignment, where the optical film has a photosensitive layer and a substrate. The exposure system directs an exposure beam from a light source to a reflective polarization modulation device. The modulated exposure beam is then reflected onto the photosensitive medium for forming a pattern onto the optical film.

130. **US 6,756,089 (US 5,928,733; US 6,242,060)**, M. Yoneya, K. Iwasaki, Y. Tomioka, H. Yokokura, K. Kondo, and Y. Nagae (**Hitachi Ltd**), *Active-matrix liquid crystal display*, C09K 19/02 (428/1.26) (June 2004).

An active-matrix liquid crystal display device having a pair of substrates at least one of which is transparent, a liquid crystal layer disposed between the pair of substrates, a group of electrodes for applying to the liquid crystal layer an electric field substantially parallel to the substrate plane and a plural number of active elements being formed on one of the pair of substrates and an alignment layer disposed between the liquid crystal layer and at least one of the pair of substrates. The alignment layer is a photo-reactive material layer, and the photo-reactive material layer is a photo-reactive alignment layer which has been subjected to linearly polarized light irradiation to selectively derive a photochemical reaction and a tilt angle of liquid crystal molecules on the alignment layer surface of the substrate is 3 degrees or less.

131. **US 6,764,724 (US 7,014,892)**, M. S. Nam and Y. S. Choi, (**LG. Philips LCD Co., Ltd**), *Alignment layer for a LCDD*, G02F 1/13, (428/1.2) (July 2004).

    The present invention relates to an alignment layer, and more particularly to an alignment layer with photo-sensitive material selected from the group consisting of a pyranose polymer, a furanose polymer, polyvinyl cinnamate, polysiloxane cinnamate, polyvinyl alcohol, polyamide, polyimide, polyamic acid and silicone dioxide.

132. **US 6,770,335**, H. H. Shin, M. S. Nam, S. H. Park, M. Ree, and S. W. Lee, (**LG. Philips LCD Co., Ltd**), *Photoalignment materials and liquid crystal display device and method for fabricating the same with said materials*, C08G 73/00 (428/1.25) (August 2004).

    Polyamideimide photoalignment materials having a photosensitive diamine derivative compound with side branches, and liquid crystal display devices using such a photoalignment material, beneficially as an alignment film.

133. **US 6,784,963**, K. H. Park, (**LG. Philips LCD Co., Ltd**), *Multi-domain liquid crystal display and method of fabricating the same*, F02F 1/13 (349/129) (August 2004).

    A multi-domain liquid crystal display and method of fabricating the same is disclosed in the present invention. More specifically, a liquid crystal display includes first and second substrates, a liquid crystal layer between the first and second substrates, wherein the liquid crystal layer a twist angle of at least 90 degrees, and an optical plate between the liquid crystal layer and the second substrate, wherein the optical plate has an optical axis horizontal to the first and second substrates.

134. **US 6,791,749**, J. DelPico and Y. G. Conturie, *Polarized exposure for web manufacture*, G02B 5/30 (359/487) (September 2004).

    A system is provided in which an expanded non-collimated source of light may be used to produce a uniform polarized light exposure for use, for example, in the photoalignment of optical films. Uniformity of polarization and intensity may be maintained even when a high-intensity source of ultraviolet light is used. The system may be scaled in size to produce large exposures without sacrificing uniformity of intensity or uniformity of direction of polarization. The system includes a light source, a pile-of-plates polarizer, and a surface (such as the surface of an optical film) to be exposed. The pile-of-plates polarizer is oriented orthogonally to the surface, thereby providing a polarized light exposure having a uniform direction of polarization on the exposed surface. The light source may be oriented at Brewster's angle to the polarizer to improve polarization contrast. Other optional features of the system are described for increasing the intensity and improving the uniformity of intensity of the polarized light exposure, and for producing polarized light exposures having different orientations on the surface.

135. **US 6,793,987 (US 7,083,833)**, H. H. Shin; H. Hyun, M. S. Nam, S. H. Park, M. Ree, and S. W. Lee (**LG. Philips LCD Co., Ltd**), *Photoalignment materials and liquid crystal display fabricated with such photoalignment materials*, C08G 69/02 (428/1.25) (September 2004).

    An aromatic polyamide photoalignment material prepared by a reaction from a diamine compound with a side branch and a photosensitive dicarboxylic acid.

136. **US 6,822,713**, O. Yaroshchuk, Y. Reznikov, J. R. Kelly, L. C. Chien, and T. Sergan, *Optical compensation film for liquid crystal display*, (**Kent State University**), G02B 5/128 (349/117) (November 2004).

    An optical compensation film for a liquid crystal display is provided wherein the film is a polymer that is capable of producing light induced anisotropy. The film is irradiated with light to form an optical axis or axes.

137. **US 6,844,913**, C. F. Leidig, (**Eastman Kodak Co.**), *Optical exposure apparatus for forming an alignment layer*, G03B 27/00 (355/2) (January 2005).

    A system for processing a multilayer liquid crystal film display material, with multiple irradiation apparatus for applying a zone of polarized UV irradiation onto a substrate fed from a web with incident light at a desired angle. Each irradiation apparatus includes a UV light source, and one or more optional filters. A polarizer is provided, sized and arranged to polarize light over the web as it moves. The irradiation apparatus employs an array of louvers and/or a prism array. One irradiation apparatus irradiates a first LPP1 layer at a 0-degree alignment, in the web movement direction, the other irradiation apparatus irradiates an LPP2 layer with an orthogonal 90-degree alignment.

138. **US 6,927,823**, Y. Reznikov, J. West, and O. Yaroschuk, (**Kent State University**), *Method for alignment of liquid crystals using irradiated liquid crystal films*, G02F 1/1337 (349/124) (August 2005).

    A method is provided for forming an alignment layer for a liquid crystal cell that is made from a liquid crystal film that has been irradiated with light. The method includes the steps of disposing a liquid crystal film on a substrate and then irradiating the liquid crystal film with light. Also, a liquid crystal display that includes an alignment layer that is a liquid crystal film that has been irradiated with light.

139. **US 6,939,587**, S. Kumar, L. C. Chen, and J. H. Kim, (**Kent State University**), *Fabrication of aligned crystal cell/film by simultaneous alignment and phase separation*, B05D 3/06 (427/510) (September 2005).

    A method for simultaneously fabricating a phase separated organic film and microstructures with liquid crystal having desired alignment is disclosed. The method includes the step of preparing a mixture of liquid crystal material, prepolymer, and polarization-sensitive material. The mixture is disposed on a substrate and a combination of UV or visible light or heat treatment is applied while simultaneously inducing phase separation so as to form a layer or microstructure of appropriately aligned liquid crystal material adjacent the substrate.

140. **US 6,943,930**, X. D. Mi, D. Kessler, R. Liang, and T. Ishikawa, (**Eastman Kodak Co.**), *Method and system for fabricating optical film using an exposure source and reflecting surface*, G02B1/10 (359/247) (September 2005).

    An exposure system for fabricating optical film such as for photoalignment, where the optical film has a photosensitive layer and a substrate. The exposure system directs an exposure beam from a light source through the optical film, then uses a reflective surface to reflect the exposure energy back through the optical film to enhance or otherwise further condition the photoreaction of the photosensitive layer.

141. **US 6,988,811 (US 6,874,899)**, C. F. Leidig, D. Kessler, J. Donner, R. Liang, and X. D. Mi, (**Eastman Kodak Co.**), *Optical exposure apparatus and method for aligning a substrate*, F21V 9/14 (362/19) (January 2006).

    A system for processing a multilayer liquid crystal film display material, with multiple irradiation apparatus for applying a zone of polarized UV irradiation onto a substrate fed from a web with incident light at a desired angle. Each irradiation apparatus includes a UV light source, and one or more optional filters. A polarizer is provided, sized and arranged to polarize light over the web as it moves. The irradiation apparatus employs an array of louvers and/or a prism array. One irradiation apparatus irradiates a first LPP 1 layer at a 0-degree alignment, in the web movement direction, the other irradiation apparatus irradiates an LPP 2 layer with an orthogonal 90-degree alignment.

142. **US 7,044,600 (US 7,097,303; US 7,097,304; US 7,101,043)**, A. Kumar and P. C. Foller, (**PPG Industries Ohio, Inc.**), *Polarizing devices and methods of making the same*, G02C 7/10 (351/163) (May 2006).

Certain, non-limiting embodiments of the disclosure provide ophthalmic elements and devices comprising an at least partial coating adapted to polarize at least transmitted radiation on at least a portion of at least one exterior surface of an ophthalmic element or substrate. Further, according to certain non-limiting embodiments, the at least partial coating adapted to polarize at least transmitted radiation comprises at least one at least partially aligned dichroic material. Other non-limiting embodiments of the disclosure provide methods of making ophthalmic elements and devices comprising forming an at least partial coating adapted to polarize at least transmitted radiation on at least a portion of at least one exterior surface of the ophthalmic element or substrate. Optical elements and devices and method of making the same are also disclosed.

143. **US 7,060,332**, S. B. Kwon, M. H. Lee, S. H. Park, Y. Reznikov, Y. Kurioz, and I. Gerus, (**LG. Philips LCD Co., Ltd**), *Liquid crystal display with alignment film of polyphenylenphthalamide-based material and method for fabricating the same*, C09K 19/02 (428/1.25) (June 2006).

An LCD and a method for fabricating the LCD includes a first substrate and a second substrate, a liquid crystal layer between the first substrate and the second substrate, and a coating of a polyphenylenephthalamide based material on at least one of the substrates, thereby providing an LCD having a reduced residual image, a strong anchoring energy, and a high thermal stability is achieved.

144. **US 7,074,344**, S. Nakata and M. Kimura, (**JSR Corporation**), *Liquid crystal aligning agent and liquid crystal display element*, C09K 19/56 (252/299.4) (July 2006).

The present invention provides a liquid crystal aligning agent which provides a liquid crystal alignment film having surface anchoring force and pretilt angle development stability by an optical aligning method. The liquid crystal aligning agent comprises a polymer having a photo-crosslinkable structure, a structure having at least one group selected from the group consisting of a fluorine-containing organic group, an alkyl group having 10 to 30 carbon atoms and alicyclic organic group having 10 to 30 carbon atoms and optionally, (C) a thermally crosslinkable structure.

145. **US 7,083,833 (US 6,793,987)**, H. H. Shin, M. S. Nam, S. H. Park, M. Ree, and S. W. Lee, (**LG. Philips LCD Co., Ltd**), *Photoalignment materials and liquid crystal display fabricated with such photoalignment materials*, G02F 1/1337 (428/1.1) (August 2006).

    An aromatic polyamide photoalignment material prepared by a reaction from a diamine compound with a side branch and a photosensitive dicarboxylic acid. Furthermore, liquid crystal display devices using such photoalignment materials in an alignment film on at least one substrate.

146. **US 7,118,787 (US 6,830,831; US 7,081,307)**, M. O'Neill, S. M. Kelly, A. E. A. Contoret, G. J. Richards, and D. Coates, (**University of Hull**), *Liquid crystal alignment layer*, H05B 33/00 (428/1.2) (October 2006).

    There is provided a liquid crystal alignment layer comprising an alignment layer; and chemically bound to said alignment layer, a transport material. Also provided are methods for forming the liquid crystal alignment layer and the use thereof in displays for electronic apparatus.

147. **US 7,221,420 (US 6,897,926; US 6,900,866; US 6,909,473; US 6,954,245; US 6,982,773; US 7,023,512; US 7,061,561; US 7,184,115)**, B. D. Silverstein, A. Kurtz, and X. D. Mi, (**Sony Corporation; Moxtek Inc.**), *Display with a wire grid polarizing beamsplitter*, G02F 1/1335 (349/117) (May 2007).

    A system for creating a patterned polarization compensator has a retardance characterization system for optically scanning the spatially variant retardance of a spatial light modulator. A compensator patterning system writes a spatially variant photoalignment pattern on a substrate of a polarization compensator. The patterned polarization compensator is completed by a process that includes providing a photoalignment layer onto which spatially variant photoalignment layer is formed, providing a liquid crystal polymer layer onto the photoalignment layer, and then fixing the liquid crystal polymer layer to form a spatially variant retardance pattern into the structure of the patterned polarization compensator.

148. **US 7,244,627**, Y. B. Lee, Y. S. Ham, S. H. Park, and S. H. Nam, (**LG. Philips LCD Co., Ltd**), *Method for fabricating liquid crystal display device*, H01L 21/00 (438/30) (July 2007).

    A method for fabricating a liquid crystal display (LCD) device to improve picture quality by preventing defective rubbing, is disclosed. The method which includes preparing first and second substrates, forming a thin film transistor on the first substrate, forming a first orientation layer on the first substrate including the thin film transistor, performing rubbing and orientation direction alignment processes on the first orientation layer to provide a uniform alignment direction, and forming a liquid crystal layer between the first and second substrates.

149. **US 7,286,275 (US 7,256,921)**, A. Kumar, P. C. Foller, F. R. Blackburn, J. Shao, M. He, and T. A. Kellar, (**Transition Optical, Inc.**), *Polarizing, photochromic devices and methods of making the same*, G02F 1/03 (359/241) (October 2007).

    Provided is an optical element including a substrate; a first at least partial coating of an at least partially ordered alignment medium connected to at least a portion of at least one surface of the substrate; a second at least partial coating of an alignment transfer material that is connected to and at least partially aligned with at least a portion of the at least partially ordered alignment medium; and a third at least partial coating connected to at least a portion of the alignment transfer material, the third coating including at least one anisotropic material that is at least partially aligned with at least a portion of the at least partially aligned alignment transfer material and at least one photochromic–dichroic compound that is at least partially aligned with at least a portion of the at least partially aligned anisotropic material.

150. **US 7,298,429**, H. Yokoyama, J. Yamamoto, and L. Komitov, (**Japan Science and Technology Agency**), *Photoinductive switching liquid crystal device*, G02F 1/135 (349/24) (November 2007).

    A fast and highly practical photoinduced switching liquid crystal device is provided. In the photoinduced switching liquid crystal device, in-plane switching of an optical anisotropic axis of dichroic nematic liquid crystal is performed at a high speed by light.

7.3 Analysis and Comments on the Patents

We try to organize the above patents in two ways: in terms of materials and in terms of companies filing the patents. Table 7.1 shows the patents classified into materials and different types of technologies. Table 7.2 groups the patents according to the assignees.

From these tables, several interesting conclusions can be drawn. The qualitative analysis of the patents shows that:

1. Photoalignment is an area of intense research interest and remains a 'hot topic' of research. The area of surface photoalignment for LC displays has reached maturity, but there is much scope and incentive for further work.
2. Photoinduced LC alignment involves the use of polarized light either to induce an easy axis in a homogeneous alignment layer or to switch from homeotropic to homogeneous surface conditions. Unpolarized light may also be used for the purpose.
3. Four main mechanisms of the photoalignment are known: (i) reversible photochemical *trans-cis* isomerization in azo-dye-containing polymers, monolayers, and pure dye films; (ii) pure photophysical reorientation of the azo-dye chromophore molecules; (iii) topochemical crosslinking in cinnamoyl side-chain polymers; (iv) photodegradation in polyimide materials.
4. Molecular design and synthesis have been used to improve significantly the performance of photoalignment layers. Photochemically and thermally stable materials are now available, which show strong LC anchoring and require low exposure threshold intensities to achieve LC alignment.
5. Pretilt alignment with pretilt angles suitable for TN-, STN- and VAN-LCD has been successfully achieved. Photoalignment is an attractive non-contact LC alignment technology and its ability for easy patterning gives it an extra advantage in the manufacturing of LCDs with multidomain pixels, e.g. for LCDs with wide viewing angles. A controllable pretilt angle and anchoring energy enable the development of a new generation of LC devices: with low voltage, fast response time, wide viewing angles, and bi- and multistable switching.
6. The photoalignment process could be used for the production of new, emerging LCD types such as ferroelectric LCD, bistable nematic LCD, LCD on

flexible substrates as well as polarized EL and OLED displays. A more general application is as high-resolution, photo-patternable optical retarders and polarizers, which would be useful in 3D and transflective LCD.

7. These fields, where photoalignment technology happens to be useful, may cover the ordering of thin semiconductor layers, thin layers in solar cells, optical data storage, and holographic memory. New, highly efficient photovoltaic, optoelectronic, and photonic devices thus become possible.

Index

Photoalignment of Liquid Crystalline Materials: Physics and Applications
V. Chigrinov, V. Kozenkov and H.-S. Kwok